SpringerBriefs in Molecular Science

SpringerBriefs in Biometals

More information about this subseries at http://www.springer.com/series/10046

Gerhard Geipel

Uranium and Plant Metabolism

Measurement and Application

Gerhard Geipel
Institute of Resource Ecology
Helmholtz-Zentrum Dresden-Rossendorf
Dresden, Sachsen, Germany

ISSN 2191-5407 ISSN 2191-5415 (electronic)
SpringerBriefs in Molecular Science
ISSN 2212-9901 ISSN 2542-467X (electronic)
SpringerBriefs in Biometals
ISBN 978-3-030-80814-3 ISBN 978-3-030-80815-0 (eBook)
https://doi.org/10.1007/978-3-030-80815-0

This Springer imprint is published by the registered company Springer Nature Switzerland AG
The registered company address is: Gewerbestrasse 11, 6330 Cham, Switzerland

Acknowledgements

The author would like to thank actual and former co-workers for their help. Special thanks to K. Viehweger, S. Sachs, G. Grambole, J. Seibt, and S. Heller.

The author would especially like to thank L Barton for his assistance during the preparation of the final manuscript.

Contents

Chapter 1
Introduction

Elements with atomic numbers >89 are named actinides or, in the case of atomic numbers >103, transactinides. All isotopes of these elements are unstable due to radioactive decay. The elements of the actinide series fill the 5f shell, meaning that their electron configuration commonly follows $[Rn]5f^{1-14}6d^{1}7s^{2}$. There are two important exceptions: Thorium has the configuration $[Rn]5f^{0}6d^{2}7s^{2}$ which rationalises its tetra valence. Berkelium has an electron configuration $[Rn]5f^{9}6d^{0}7s^{2}$ and so favors oxidation state +2. The restricted availability of the actinides with atomic numbers higher than 97 (Berkelium) plus their short half-life result in a relatively low interest in these elements in environmental research.

The uranium element belongs the actinide series. According to its place in the periodic table, uranium is a member of the lower actinides. Actinides are distinguished into naturally-occurring and synthetic elements. A sharp line between synthetic and naturally occurring elements cannot be drawn. Synthetic uranium isotopes are produced by human activities in using and testing nuclear power and nuclear weapons. In smaller degree due to natural nuclear reactors in the Proterozoic era, plutonium is among non-synthetic elements. Therefore, thorium, protactinium, uranium and, in much smaller amounts, plutonium occur under natural conditions. All other actinides including neptunium are synthetic elements. The actinide elements uranium, neptunium, plutonium as well as americium show different oxidation states. Due to this property, these elements show redox reactions and this may be of great significance in biological systems.

All actinides are non-essential elements. They are radioactive and show decay properties and are heavy metals. Therefore, actinides are counting among high hazardous elements. The natural occurring elements uranium, thorium are mother nuclides of decay series. The transuranium elements neptunium, plutonium, americium and curium were mainly generated in nuclear power stations: Therefore, all these elements have to be considered in nuclear waste management and nuclear accidents.

G. Geipel, *Uranium and Plant Metabolism*, SpringerBriefs in Biometals,
https://doi.org/10.1007/978-3-030-80815-0_1

The toxicity of an element depends strongly on its chemical speciation. This influences also the bioavailability of these elements. In case of uranium, its toxicity decreases in the series:

uranyl phosphates > uranyl citrates > uranyl carbonates.

Knowledge about the speciation of actinides under different natural conditions is therefore of great importance.

Uranium, americium and curium show luminescence properties. Most studies, using this property, been performed with uranium. Two reasons for this should be named. Uranium is the last element in the periodic table occurring naturally in higher amounts and therefore available in higher amounts for chemical and biological experiments in the laboratory. On the other hand, the amount of uranium needed for luminescence experiments can be handled somewhat easier other actinides. This is due to their high specific radioactivity of these elements and special equipment for safety and radiation protection, like glove boxes, has to be installed. For example, americium has strong gamma emission, which can require additionally lead shielding of the experiment. Radioelements, like neptunium, plutonium, americium and curium are much less studied than others. This is due to the difficult handling of the radioelements. They are all α-emitting radionuclides. Special equipment in the laboratories is necessary as well as regulations of radiation protection have to be considered.

Within the food chain, plants play a very important part. Knowledge about the speciation and the metabolism of actinides in the several plant cells may help to avoid the uptake of these elements and generate hazards to living organism and point ways to protect living beings.

Up to now knowledge about actinide speciation in organisms including plants is very rare. Many publications up to now deal only with so-called transfer factors. Some publications show that phosphate species often play an important role. On the other hand, during the uptake intermediates of other binding forms may become important. The hazardous potential of these actinide species may be estimated if information about their binding forms especially in storage compartments inside the plant cells is available. The luminescence properties of the named elements can be used for the determination of the binding forms.

The most impact actinides in biological systems is connected to the release of radioactive elements in nuclear accidents as Chernobyl and Fukushima. Also, scenarios which are seen in the discussion of radioactive waste storage play an important role.

Microorganism are the first organism which may have contact to actinides in radioactive waste storage sites. After transport to the earth surface these radionuclides may also access the food chain via several ways of uptake. Among these is also the uptake of actinides by plants.

However, it depends strongly on the bioavailability of these elements and thereby on the speciation. The speciation may also change during and after the uptake.

Table 1.1 Actinide isotopes with the longest half-life and oxidation states

Atomic number	Element symbol	Isotope	Half life	Main emitter	Oxidation states
89	Ac	227	21.7y	β^-	***+3***
90	Th	232	14E09y	α	***+4***
91	Pa	231	32,500y	β^-	+3; +4; ***+5***;
92	U	238	4.4E09y	α	+3; +4; +5; ***+6***
93	Np	237	2.14E06y	α	+3; +4; ***+5***; +6; +7
94	Pu	244	81E06y	α	+2; +3; +4; +5; + 6; +7
95	Am	243	7340y	α	+2; ***+3***; +4; +5; + 6
96	Cm	247	15.6E06y	α	+2; ***+3***; +4
97	Bk	247	1400y	α	***+2***; +3; +4
98	Cf	251	900y	α	+2; +3; +4
99	Es	252	472d	α; β^+	+2; +3; +4
100	Fm	257	100.5d	α	+2; +3; +4
101	Md	258	51.5d	α	+2; +3
102	No	259	58 min	α; β^+	+2; +3; +4
103	Lw	262	3.6 h	β^+	+3

(Main oxidation state in bold and italic face)

A broad review about actinides in biological systems, especially in animals and man is given by P.W. Durbin in "The Chemistry of Actinide and Transactinide Elements" [1]. It serves as an excellent source of work published before 2005.

An introduction in the uptake of radionuclides by plants has been given by Greger [2]. Besides the description of common uptake mechanisms only little information about actinides is given.

An overview about soil to plant transfer is given by Robertson et al. [3]. The factors vary very strong depending on plant species, soil and experimental conditions, but compared to Sr isotopes the values for actinides are smaller.

To give an insight to the actinides chemistry all actinides in their most stable with half-life, radioactive decay and main oxidation states are summarized in Table 1.1.

Several actinides show luminescence properties which enable their study by emission spectroscopic methods. Besides uranium, which is luminescent in all environmentally relevant oxidation states, americium and curium, especially, can be determined by time-resolved laser-induced fluorescence spectroscopy. Measurement of luminescence lifetimes of species of these two elements enables determination of their hydration number [4]. The wide-ranging isotopic variety of actinides can be appreciated in the nuclide table. The most stable nuclides which may be relevant to biological systems are summarized in Table 1.1. More details may be found in special nuclide charts [5, 6].

The availability of transuranium elements in marine environments was surveyed by P. Scoppa in 1984 [7]. It was estimated that, by the year 2000, about 288 metric tons of man-made transuranium elements had arrived in marine environments [8]. The highest amount should be ^{237}Np (~194 tons or about 68%) followed by ^{243}Am (38.9 tons or 13.6%).

Radionuclide concentrations of some actinides have been measured in benthic invertebrates (rock jingle, blue mussel and horse mussel) in the islands of the Aleutian Chain [8]. The highest levels detected were for ^{234}U (0.45–0.84 Bq/kg wet weight). The concentrations are slightly smaller for ^{238}U whereas those for ^{241}Am, ^{239,240}Pu and ^{235}U are about one order of magnitude lower. The concentration was close to the detection limit for ^{236}U. The presence of isotopes ^{241}Am, ^{239,240}Pu and ^{236}U are due to human activities in the environment.

Several studies deal with the rapid determination of actinides in different samples, mainly human urine. By the use of inductively coupled mass spectrometry and alpha spectrometry in combination with a rapid separation technique, nearly all actinide isotopes can be determined in a relatively short period of time [9, 10].

References

1. Durbin PW (2006) In: Morss LR, Edelstein NM, Fuger J (eds) The chemistry of actinide and transactinide elements, 3rd edn, vol 5, Springer New York 3339–3440
2. Greger M (2004) Uptake of nuclides by plants. SKB Technical report TR-04-14, Stockholm
3. Robertson DE, Cataldo DA, Napier BA (2003) Literature review and assessment of plant and animal transfer factors used in performance assessment modeling, NUREG/CR-6825 PNNL-14321
4. Collins RN, Saito T, Aoyagi N, Payne TE, Kimura T, Waite TD (2011) Applications of time-resolved laser fluorescence spectroscopy to the environmental biogeochemistry of actinides. J Env Qual 50:731–741
5. Pfennig G, Klewe-Nebenius H, Seelmann-Eggebert W (1998) Karlsruher Nuklidkarte, Forschungszentrum Karlsruhe
6. www.nndc.bnl.gov/nudat2/
7. Scoppa P (1984) Environmental behavior of trans-uranium actinides - availability to marine biota. Inorg Chim Acta 95:23–27
8. Burger J, Gochfeld M, Jeitner C, Gray M, Shukla T, Shukla S, Burke S (2007) Radionuclide concentrations in benthic invertebrates from amchitka and kiska islands in the Aleutian Chain. Alaska Environ Monit Assess 128:329–341
9. Maxwell SL (2008) Rapid analysis of emergency urine and water samples. J Radioanal Nucl Chem 275:497–502
10. Maxwell SL, Jones VD (2009) Rapid determination of actinides in urine by inductively coupled plasma mass spectrometry and alpha spectrometry: a hybrid approach. Talanta 80:143–150

Chapter 2
Modern Methods of Uranium Detection

To determine uranium several methods are applicable. The total uranium content may mostly determined by ICP-MS measurements. Of more interest are methods, which allow to assign the uranium speciation in the environment under investigation.

Normal UV–Vis measurements have detection limits above environmental concentrations ranges. Therefore, mostly methods with high intense light sources are used to determine uranium speciation under these conditions.

Other methods which have great potential for biosystems are under development. As example the secondary neutral ionisation SIMS should be named.

2.1 Laser-Induced Spectroscopy

Laser-induced spectroscopic methods decrease the detection limits up to 5–7 orders of magnitude, depending on the method of detection and on spectroscopic properties of the studied element. By use of common UV–Vis measurements uranium(VI) can be measured up to about 10^{-2} M solutions (without special colouring reagents). Laser induced absorption spectroscopy (so-called Laser-induced photoacoustic spectroscopy, LIPAS) reaches already detection limits of 10^{-5} M. Last not least for uranium also luminescence properties can be used for the proof of this element, reaching in special cases (Cryo-TRLFS, carbonate complexes) detection limits of 2×10^{-10} M. The principal fundamental processes in a sample after application of a laser pulse are shown in Fig. 2.1.

In the field of laser-induced spectroscopy five main methods are used: laser-induced photo- acoustic Spectroscopy (LIPAS), Thermal lensing spectroscopy (TLS), laser-induced time-resolved fluorescence spectroscopy (TRLIFS), Ultra-short laser pulse-induced time-resolved fluorescence spectroscopy (fs-TRLIFS), laser-induced breakdown detection and spectroscopy (LIBD/LIBS). All of these methods have been intensively developed during the last decades [1–9]. They became powerful tools to study interactions in solutions and at the solid–liquid interface. LIBD/LIBS

G. Geipel, *Uranium and Plant Metabolism*, SpringerBriefs in Biometals,
https://doi.org/10.1007/978-3-030-80815-0_2

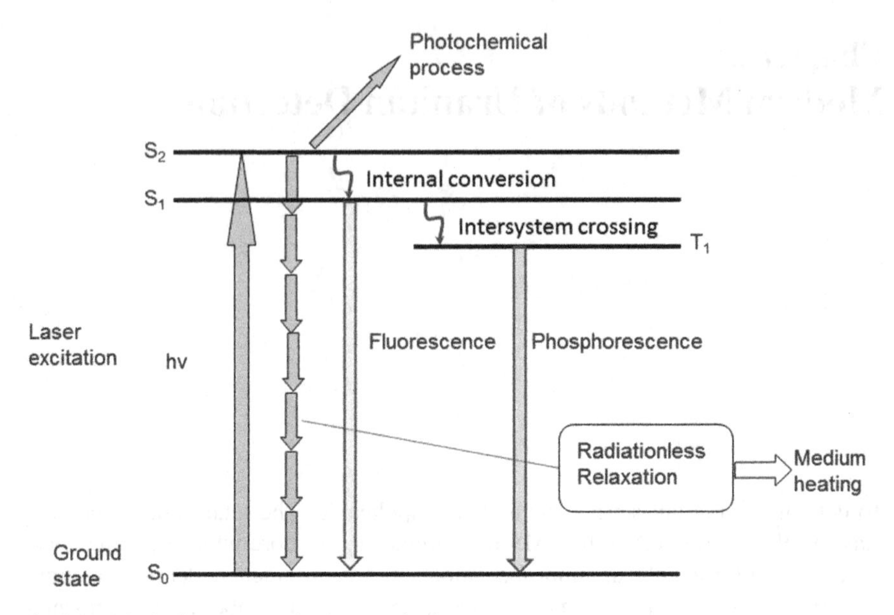

Fig. 2.1. Principal processes after laser excitation

concerns especially studies in colloidal systems. This theme should be the aim of another contribution especially under the point of view that the sample will we changed due to the destruction of particles. Therefore, it would be pardonable that LIBD is not treated here.

Main field in speciation techniques using lasers are lanthanide and actinide chemistry. While lanthanide often have been used as probe system for energy transfer studies in biochemistry and organic chemistry, the speciation studies of actinides focus directly on the behavior of these elements, especially in environmental systems as uranium mining areas and waste disposal sites.

The possibility of two photon excitation has been offered with reason of the development of pico-second pulse and femto-second pulse lasers. Especially for sensitive organic and biologic samples this provides opportunities for mild excitation [10].

All laser induced methods need a computer system for controlling the experiment and for data storage. For fluorescence measurements commercial program codes are available.

2.1.1 Laser-Induced Photoacoustic Spectroscopy (LIPAS)

Photoacoustic spectroscopy employs the so-called photoacoustic effect discovered by G. Bell already in 1880. It employs the generation of sound by absorption of modulated light.

The modulated light can be generated in two ways:

- a cw beam is chopped typically with 10–1000 Hz or
- a pulsed light beam with a duration <1 μs is used

This allows a gated (boxcar) or time resolved measurement.

The advantages of the method are that only sound will be produced by light illumination in the measured liquid and ambient noises can be discriminated by bandpass filters. For detection of the generated acoustic wave by pulsed light sources, piezoelectric transducers with fast rise time are ideal.

A detailed description of the basics of this method can be found in Tam and Patel [11]. The basic equation for LIPAS is shown below.

$$\frac{a}{\mathrm{E}} = \frac{\beta}{2\pi r C_p \rho}\alpha \tag{2.1}$$

α absorption coefficient
β thermal coefficient of expansion
ρ density of the liquid
r radius of the illuminated liquid column
a change of the radius caused by the laser pulse
C_p specific heat at constant pressure
E energy of the laser pulse.

Figure 2.2 shows a setup for laser-induced photoacoustic spectroscopy. The pulses of a tunable laser system (dye laser or optical parametrical oscillator system) are directed into a cuvette with the solution under investigation. A small amount of the laser energy at the desired wavelength will be absorbed. After going away of the laser pulse, the absorbed energy will be released to the solution as the excitation is

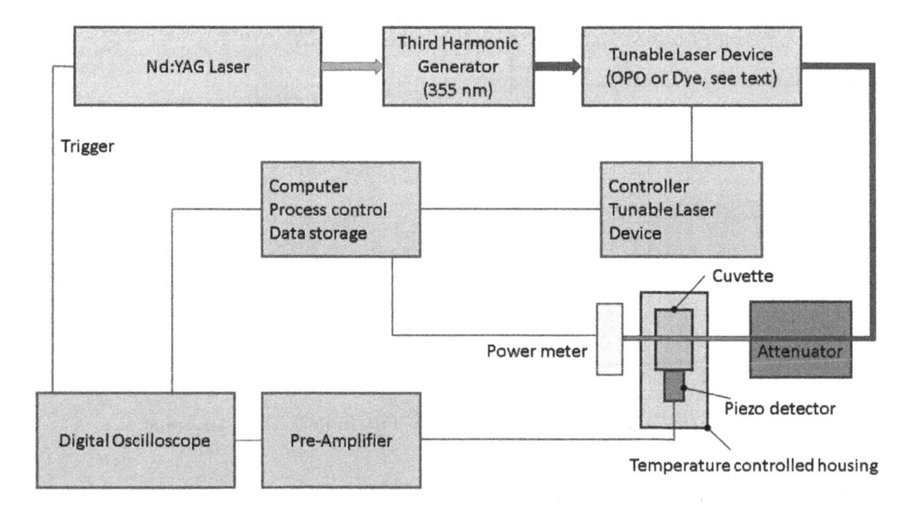

Fig. 2.2. Simplified setup of laser-induced photoacoustic spectroscopy

going to the ground state. The energy release occurs only in the small volume where the laser pulse was applied. The outer volume of the solution acts as a thick wall. So the released energy produces pressure wave via a temperature jump in the volume (expansion of the illuminated volume). The pressure wave can be detected using piezo ceramic transducers as an electrical signal. The amount of the signal depends directly on the absorption of laser light and the applied laser energy. Nevertheless, it has to be taken into account that the applied laser energy must be below the breakdown limit of solutions (about 5 mJ). By dividing with the applied energy, a value is received which corresponds to the absorption of the solution at the desired wavelength. Scanning over the wavelength range of interest a spectrum will be obtained comparable to classic absorption spectra. The advantages of this method are:

- direct measurement of the absorption
- high power of the applied light.

This leads to a detection limit about 2–3 orders of magnitude lower than for conventional UV–vis spectrometry. However, the installation of this method is very expensive. Therefore, the application is very limited and is only applied in actinide chemistry where the radioactivity limits often the handling of higher concentrations of the elements.

A typical photoacoustic signal wave (oscilloscope picture) is shown in Fig. 2.3. The trigger point is marked with "T", this means the time, when the laser pulse hits the cuvette. About 30 μs after the laser pulse the start of the photoacoustic wave can be observed. This delay is due to the transition time of the photoacoustic pressure wave between the location of the laser beam in the cuvette and the piezo-ceramic

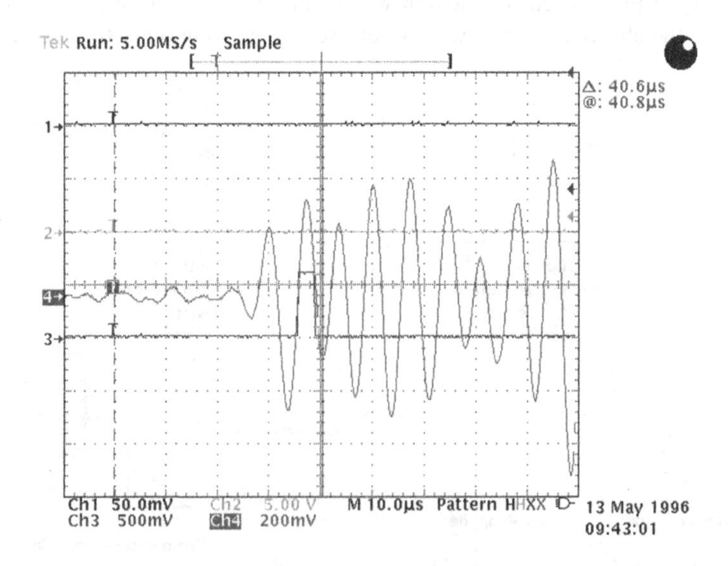

Fig. 2.3. Oscilloscope picture of a photoacoustic wave

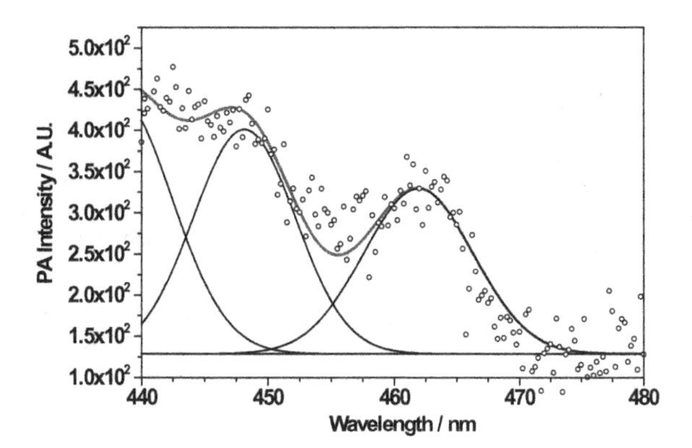

Fig. 2.4. Photoacoustic absorption spectrum of uranium from a mine tailing water

transducer located at the wall of the cuvette. The delay time depends on the length of this travel and the speed of the wave in the medium.

As result one get a so called photoacoustic absorption spectrum, which is adequate to a normal absorption spectrum. A section of the photoacoustic spectrum of a tailing water from a former uranium mining site is shown in Fig. 2.4. The water contains 2.5 μmol/L uranium(VI) and 10.3 mmol/L HCO_3^- and the pH was 9.76. The spectrum could be clearly assigned as $UO_2(CO_3)_3^{4-}$ [7].

2.1.2 Time-Resolved Laser-Induced Fluorescence Spectroscopy (TRLIFS)

The third and most common method (besides LIPAS and TLS) in application of lasers in direct speciation techniques is the laser-induced time-resolved fluorescence spectroscopy (TRLIFS). However, the method is limited to luminescent metal ions of the lanthanide and actinide series as europium, uranium curium. Figure 2.5 shows a common setup of the method. The tunable laser device should generate laser pulses below 470 nm if used in uranium(VI) systems. The laser pulse is applied to the sample and in a rectangular setup the emitted luminescence is focused into the entrance of the spectrograph. The deconvoluted spectrum is measured by use of a so-called intensified CCD (charged coupled device) camera. More details are described in several publications [6, 12–15]. Due to the need to determine complete spectra other detection methods as single photon counting are much less exploited in speciation techniques [16].

The luminescence properties of lanthanides have been widely used in biochemistry especially exploiting energy transfer effects.

The development of laser systems in direction of tunable lasers and short pulse length and different detection possibilities lead to observation of up to now unknown luminescence properties of some actinide elements in aquatic environments. Besides

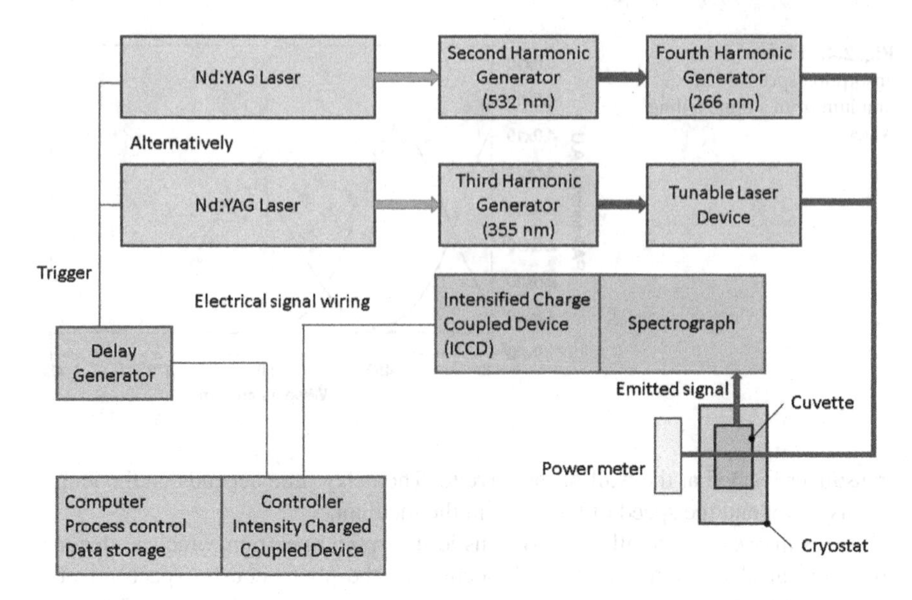

Fig. 2.5. Setup for time resolved laser induced fluorescence spectroscopy (TRLFS) using a tunable laser system designed for UO_2^{2+} measurements

the detection of uranium(V) luminescence [17] also neptunium(VI) emission in the near infrared are possible.

Several laser systems are be suggested to provide different excitation wavelength. Most common tunable laser systems are dye lasers pumped by an excimer laser or Nd:YAG laser. About 30 years ago also tunable solid-state lasers (Nd:YAG pumped optical parametric oscillators) have been established. These laser systems are able to provide tunable laser pulses over a wide wavelength range. They are also more stable in direction of the stability of the medium. In addition, the solid-state laser system does not have the problems of dyes (toxicity, cancer stimulating. Dye laser systems have short wavelength ranges of about 30 nm without changing the dye.

An alternative system for TRLFS measurements is shown in Fig. 2.6. This system has been used to detected the fluorescence of organic ligands. Due to short luminescence lifetimes the used ICCD-Camera system is a special system, designed for short gate opening and small delay ranges. However, the spectra of the ligands are not influenced in their emission wavelength due to interaction with metal ions. Therefore, this method cannot be used for direct speciation measurements. However, applications to study the interaction of non-fluorescent metal ions with organic fluorescent ligands can be studied with such systems [8, 18].

By use of TRLFS, two important species-specific data can be extracted: emission wavelength and luminescence lifetime. Figure 2.7 shows a typical time resolved luminescence spectrum of uranium(VI) in 0.5 M $HClO_4$. The uranium concentration was 5×10^{-5} Mol. However, this is not yet close to the detection limit. The time axis (delay after laser pulse) is not a linear one. The emission wavelength can be derived

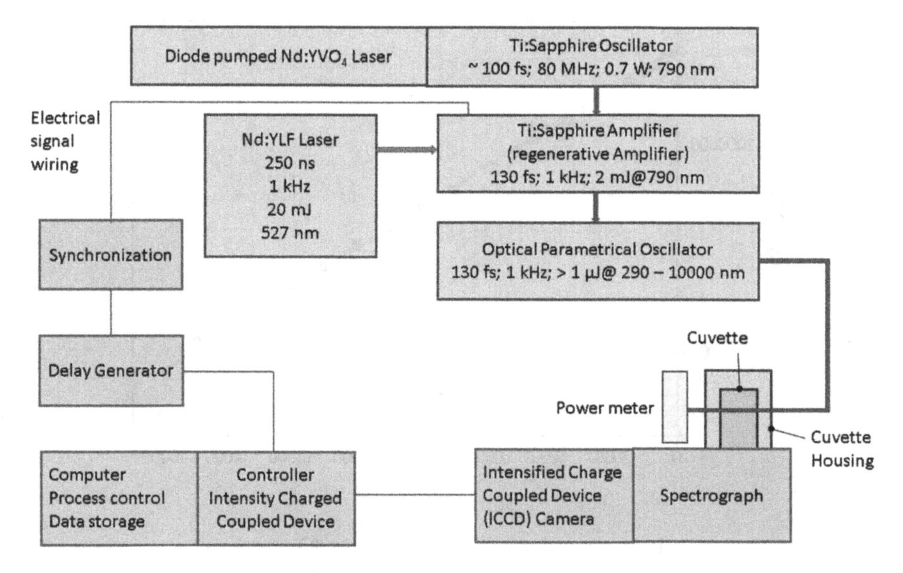

Fig. 2.6. Setup of a TRLIFS system using fs-laser pulses for excitation of molecules with fluorescence emissions and lifetimes <20 ns

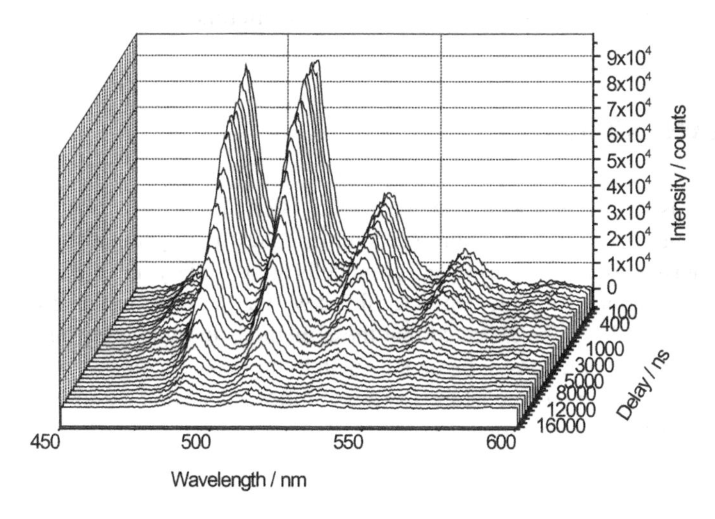

Fig. 2.7. Time-resolved emission spectrum of uranium(VI)

from a single emission spectrum were located at 472.3 nm, 488.5 nm, 509.9 nm, 533.4 nm, 559.4 nm and 580.1 nm, respectively. The emission at 472.3 nm is a hot band to the 488.5 nm emission, whereas the emissions at 509.9 nm and higher wavelength result from transitions from the 488.5 nm level to the several vibronic levels in the ground state. The associated decay curve is shown in Fig. 2.8. For linearization of the curve the intensity is in logarithmic scale. The lifetime of the

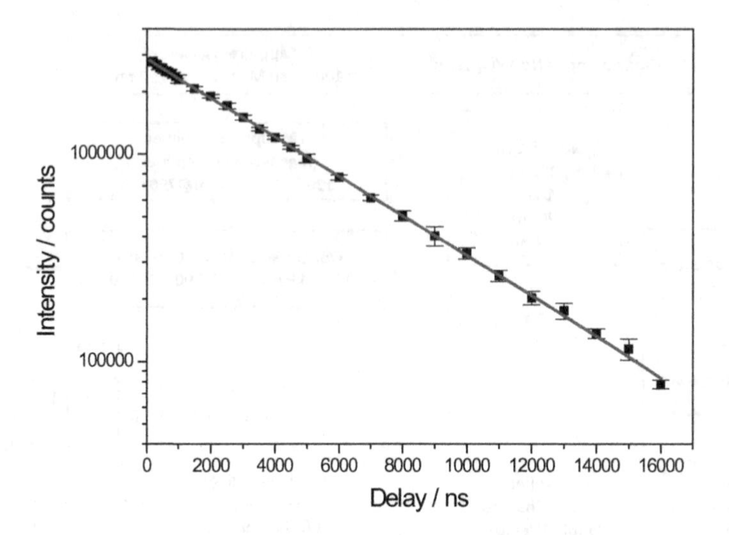

Fig. 2.8. Luminescence decay curve of uranyl ions in 0.5 M $HClO_4$

uranyl phosphorescence from this measurement in 0.5 M $HClO_4$ is assigned to be 4.6 ± 0.1 μs (for comparison in 0.1. M $HClO_4$ the lifetime is about 1.8 μs).

2.2 XANES/EXAFS

Modern tools for the localization and characterization of processes involving metals have been suggested [19]. The authors recommend the use of synchrotron techniques for the determination of metals inside plant cells as subsidiary method. XAS-based methods are automatically element-specific as well as sensitive to the environment of the metal ion. It can be expected that these methods in future play an important role to study the interactions of metal ions with living cells. Studies with bioligands, like DNA, and uranium demonstrate the possibilities and advantages of this method to obtain structure information [20].

References

1. Beitz JV, Hessler JP (1980) Oxidation-state specific detection of transuranic ions in solution. Nuc Technol 51:169–177
2. Schrepp W, Stumpe R, Kim JI, Walther H (1983) Oxidation-state specific detection of uranium in aqueous-solution by photo-acoustic spectroscopy. Appl Phys B 32:207–209
3. Kim JI (1986) In: Freeman AJ, Keller C (eds) Handbook of the physics and chemistry of the actinides, vol 4, Elsevier Science Publishers B.V., 413

4. Beitz JV, Bowers DL, Doxtander MM, Maroni VA, Reed DT (1988) Detection and speciation of transuranium elements in synthetic groundwater via pulsed-laser excitation. Radiochim Acta 44(45):87–93
5. Kim JI, Stumpe R, Klenze R (1990) Laser-induced photoacoustic-spectroscopy for the speciation of transuranic elements in natural aquatic systems. Topics Curr Chem 157:129–179
6. Klenze R, Kim JI, Wimmer H (1991) Speciation of aquatic actinide ions by pulsed laser spectroscopy. Radiochim Acta 52(53):97–103
7. Geipel G, Bernhard G, Brendler V, Nitsche H (1998) Complex formation between UO_2^{2+} and CO_3^{2-}: studied by laser-induced photoacoustic spectroscopy (LIPAS). Radiochim Acta 82:59–62
8. Geipel G, Acker M, Vulpius D, Bernhard G, Nitsche H, Fanghänel Th (2004) An ultrafast time-resolved fluorescence spectroscopy system for metal ion complexation studies with organic ligands. Spectrochim Acta A 60:417–424
9. Geipel G (2006) Some aspects of actinide speciation by laser-induced spectroscopy. Coordin Chem Rev 250:844–854
10. Piszczek G, Maliwal BP, Gryczynski I, Dattelbaum J, Lakowicz JR (2001) Multiphoton ligand-enhanced excitation of lanthanides. J Fluoresc 11:101–107
11. Tam AC, Patel CKN (1979) Optical absorptions of light and heavy water by Laser optoacoustic spectroscopy. Appl Optics 18:3348–3358
12. Billard I (2003) Handbook on the physics and chemistry of rare earths, vol 33. Elsevier Science, 465–513
13. Bidoglio G, Grenthe I, Qi P, Robouch P, Omonetto N (1991) Complexation of Eu and Tb with fulvic-acids as studied by time-resolved laser-induced fluorescence. Talanta 38:999–1008
14. Kim JI, Klenze R, Wimmer H (1991) Fluorescence spectroscopy of curium(III) and application. Eur J Sol State Inor 28:347–356
15. Aoyagi N, Toraishi T, Geipel G, Hotokezaka H, Nagasaki S, Tanaka S (2004) Fluorescence characteristics of complex formation of europium(III)-salicylate. Radiochim Acta 92:589–593
16. Simonin JP, Billard I, Hendrawan H, Bernard O, Lützenkirchen K, Semon L (2003) Study of kinetic electrolyte effects on a fast reaction in solution: the quenching of fluorescence of uranyl ion up to high electrolyte concentration. Phys Chem Chem Phys 5:520–527
17. Steudtner R, Arnold T, Großmann K, Geipel G, Brendler V (2006) Luminescence spectrum of uranyl(V) in 2-propanol perchlorate solution. Inorg Chemi Commun 9:939–941
18. Vulpius D, Geipel G, Baraniak L, Bernhard G (2006) Complex formation of neptunium(V) with 4-hydroxy-3-methoxybenzoic acid studied by time-resolved laser-induced fluorescence spectroscopy with ultra-short laser pulses. Spectrochim Acta A 63:603–608
19. Donner E, Punshon T, Guerinot ML et al (2012) Functional characterisation of metal(loid) processes in planta through the integration of synchrotron techniques and plant molecular biology. Anal Bioanal Chem 402:3287–3298
20. Rossberg A, Barkleit A, Tsushima S, Scheinost AC, Kaden P, Stumpf T (2017) A multi-method approach for the investigation of complex actinide systems: uranium(VI) interactions with DNA and sugar phosphates. Migration Conference, Barcelona, Spain

Chapter 3
Chemistry of Uranium

M. H. Klaproth discovered uranium in 1789. The pure metal was isolated in 1841 by E.-M. Peligot. Naturally occurring uranium contains the isotopes ^{238}U (99.2830% by weight), ^{235}U (0.7110%), and ^{234}U (0.0054%). ^{238}U has a half-life of 4.4E09 years. The decay chain of ^{238}U is shown in Fig. 3.1. The decay chains for natural occurring ^{232}Th can be found in the literature [1]. The chains demonstrate that the daughter nuclides of ^{238}U except ^{206}Pb are also naturally occurring radioactive isotopes. The second primary uranium isotope in the nature is ^{235}U. In Fig. 3.2 the decay chain for this isotope is also shown. In the decay chain of ^{235}U, the minor daughter decays of less than 1% are not shown. In addition, it should be mentioned, that ^{235}U is a fissile nuclide used in nuclear power stations. The abundance of uranium on earth is considered to be in the same order as molybdenum or arsenic [2]. As uranium is a ubiquitous element, its level in human and animal tissues can be about that of chromium. Up to now, 27 uranium isotopes are known, with half-life ranging from <1 μs (^{221}U) to $4.468*10^9$ a (^{238}U). Table 3.1 lists the most important (long living) isotopes and Table 3.2 summarizes all other known uranium isotopes.

Despite these isotopes are also 3 so called meta isotopes known:

$$^{218m}U\ (0.56\ ms);\ ^{235m}U\ (26\ min)\ \text{and}\ ^{238m}U\ (260\ ns)$$

Decay energies not listed in Table 3.3 were not found in the literature [3]. However, in case of the lower heavy α-decaying uranium isotopes, one can assume that the decay energies increase with decreasing atomic weight.

Depleted uranium (DU) is a by-product in the uranium enrichment process in which natural uranium (0.7% of the fissile isotope ^{235}U) is enriched. The depleted uranium contains about 0.3% ^{235}U and its behaviour in biological environments has been examined lately. New results show that the inherent coordination chemistry of uranium leads to changes in the nature of the coordination sphere when DU is transported to different biological compartments.

Uranium forms many minerals; most of them show bright yellow colors and have luminescent properties. Up to now about 190 uranium minerals are known. Of

G. Geipel, *Uranium and Plant Metabolism*, SpringerBriefs in Biometals,
https://doi.org/10.1007/978-3-030-80815-0_3

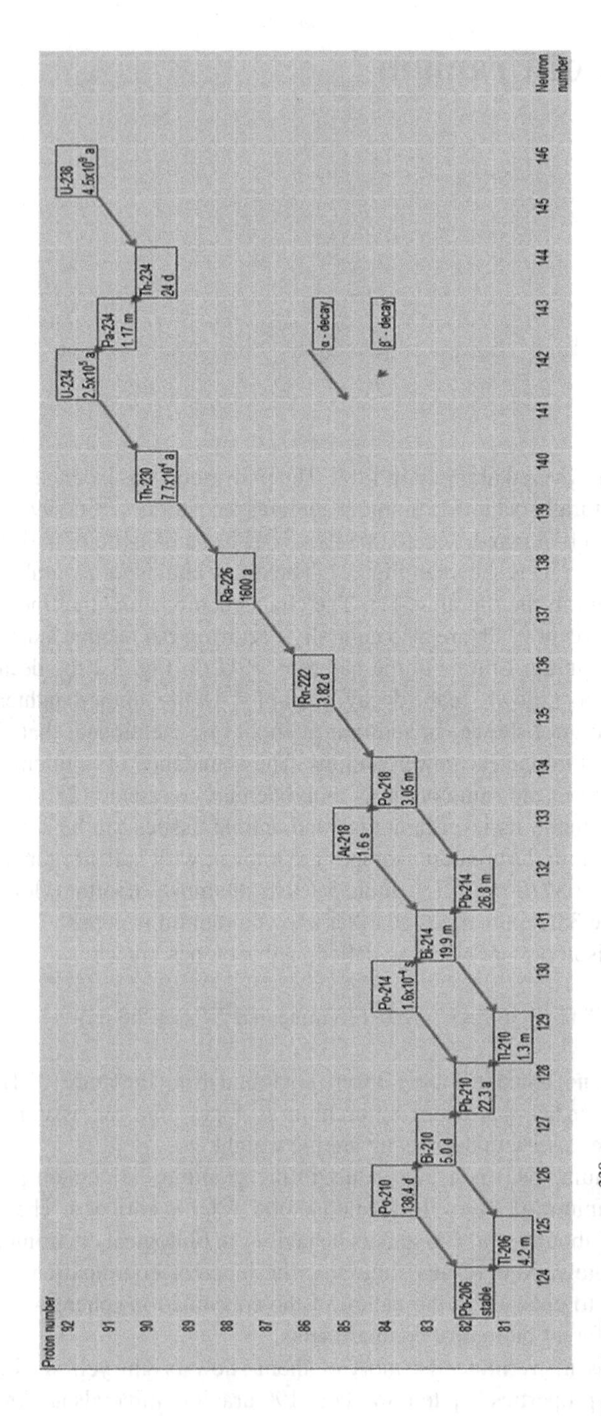

Fig. 3.1. Decay chain of ^{238}U

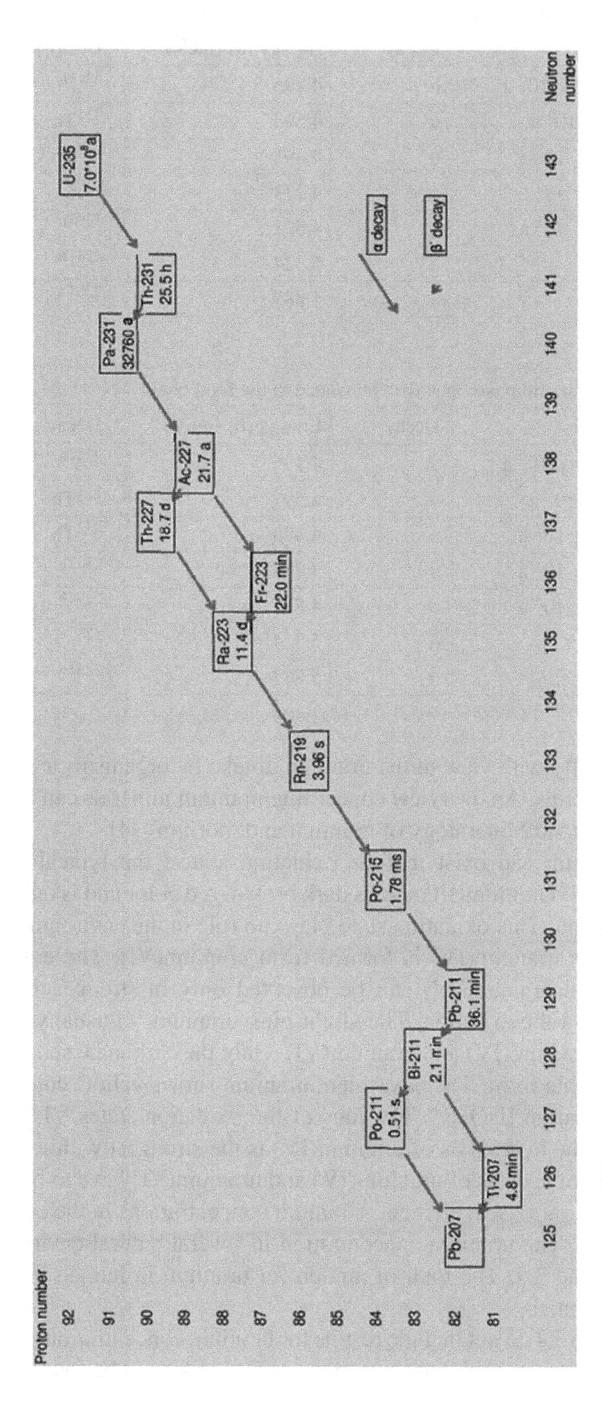

Fig. 3.2. Decay chain of ^{235}U

Table 3.1 Long lived uranium isotopes with relevance to the food chain

Isotope	Half-life	Decay	Energy (d) (meV)	Decay product
^{238}U	4.0.468 × 10^9 a	α	4.195	234Th
^{235}U	7.04 × 10^8 a	α	4.597	^{231}Th
^{236}U	2.342 × 10^7 a	α	4.493	^{232}Th
^{234}U	2.455 × 10^5 a	α	4.773	230Th
^{233}U	1.592 × 10^5 a	α	4.821	229Th
^{232}U	68.9 a	α	5.324	228Th
^{230}U	20.8 d	α	5.887	226Th

Table 3.2 Long lived uranium isotopes with relevance to the food chain

Isotope	Half-life	Decay	Energy (d) (meV)	Decay product
^{238}U	4.0.468 × 10^9 a	α	4.195	234Th
^{235}U	7.04 × 10^8 a	α	4.597	^{231}Th
^{236}U	2.342 × 10^7 a	α	4.493	^{232}Th
^{234}U	2.455 × 10^5 a	α	4.773	230Th
^{233}U	1.592 × 10^5 a	α	4.821	229Th
^{232}U	68.9 a	α	5.324	228Th
^{230}U	20.8 d	α	5.887	226Th

importance especially with view to the uranium uptake by organisms are carbonate and phosphate minerals. An overview concerning uranium minerals can be found in C. Frondel "Systematic Mineralogy of uranium and thorium" [4].

Dissolved uranium can exist in four oxidation states; the typical colors are presented in Fig. 3.3. Uranium(III) shows dark brown–red color and is only stable in non-aqueous solution. This oxidation state plays no role in the environment. Under reducing conditions uranium(IV) is formed from uranium(VI). The color is dark green and dissolved uranium(IV) can be observed only in strong acid solution. Uranium(V) forms a dioxo cation. The slight pink uranium(V) usually disproportionate rapidly to uranium(IV) and uranium(VI). Only the carbonate species a some somewhat more stable form. The hexavalent uranium shows yellow color. It forms normally a dioxo cation $[UO_2]^{2+}$. The ions of the oxidation states VI, IV and III hydrolyze easily. The hydrolysis of uranium(IV) is the strongest within this series. Aqueous solutions of species of uranium(IV) and uranium(VI) have to be acidified.

Uranium is an omnipresent element. Uranium concentrations in the environment cover a wide range. The uranium concentration in several natural environments is summarized in Table 3.4. The total of amount of uranium in human body is also given for comparison.

The data in Table 3.4 do not include results for uranium concentrations in uranium mining areas. In the waste rock concentrations of 50 mg/kg can be found.

Table 3.3 Short lived uranium isotopes

Isotope	Half-life	Decay	Energy (d) (meV)	Decay product
^{230}U	20.8 d	α	5.89	226Th
^{237}U	6.75 d	β^-	0.52	^{237}Np
^{231}U	4.2 d	α	5.46	227Th
^{240}U	14.1 h	β^-	0.4	^{240}Np
^{229}U	58 min	α	6.36	225Th
^{239}U	23.4 min	β^-	1.21	^{239}Np
^{242}U	16.8 min	β^-		^{242}Np
^{228}U	9.1 min	α	6.69	224Th
^{227}U	1.1 min	α	6.8	^{223}Th
^{226}U	350 ms	α		^{222}Th
^{225}U	69 ms	α		^{221}Th
^{224}U	900 μs	α		220Th
^{223}U	18 μs	α		219Th
^{222}U	1 μs	α		218Th
^{221}U	700 ns	α		217Th
^{219}U	42 μs	α		215Th
^{218}U	510 μs	α		214Th
^{217}U	16 ms	α		213Th

The solubility of uranium is important in relation to the bioavailability of this element. Especially formed carbonate species show high solubilities. In contrast crystalline uranium dioxide, which is formed at low pH values (pH ~ 1) shows much lower solubility then amorphous UO_2, which is formed at pH ~ 3. Solubility constants were found to be log $K_{sp}{}^0 = -59.1 \pm 1.0$ and log $K_{sp}{}^0 = -54.1 \pm 1.0$ for UO_2(cr) and $UO_2 \cdot xH_2O$(am), respectively [13]. The low solubility of uranium(IV) causes the existence dissolved uranium(IV) forms only under strong acid conditions. Uranium(VI) is much more soluble. Uranium salts like uranium(VI)-nitrate and uranium(VI)-acetate show relatively high solubility. Under alkaline pH conditions, formed carbonate-complexes show an increasing solubility. It was found that formed alkaline earth uranium(VI) carbonate species have solubilities in the range of 10 g/L.

The most important oxidation states of uranium under natural conditions are +4; + 5 and +6. A short description of important properties of these oxidation states should be given. The stable oxidation state is +6 under normal environmental conditions. However, such highly charged ions in solution do not exist, therefor uranium occurs as dioxouranium(VI) ion or with common name uranyl ion. Under reducing conditions +4 is the stable form. Uranium(IV) shows not the dioxo cationic form and hydrolyses

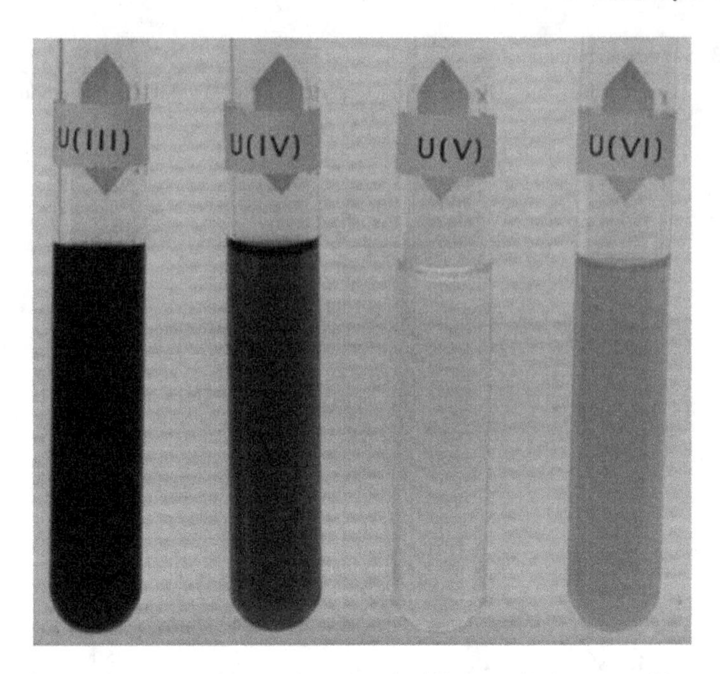

Fig. 3.3. Colors of aqueous solutions of uranium in different oxidation states; Photographs by David E. Hobart, courtesy of Los Alamos National Laboratory

Table 3.4 Range of uranium concentrations in several natural environments

Compartment	Uranium content	References
Earth's crust	2.2–4.0 mg/kg	[5]
Granitic rocks	3.0–4.02 mg/kg	[6]
Clay	2.7–5.0 mg/kg	[6]
Soil	0.5–5.0 mg/kg	[6]
Phosphate fertiliser	50–200 mg/kg	[7]
Coal	1.0–1000 mg/kg	[8]
Ocean water	1.15–3.5 μg/L	[9]
Mineral waters	<2 ng/L–188.8 μg/L	[10]
Drinking waters	<1 ng/L–73 μg/L	[11]
Air	0.02–0.1 ng/m^3	[12]
Human body (total)	20–90 μg	[12]

very easily. The solvated uranium(IV) exists only in strong acid solutions of non-complexing mineral acids like perchloric acid or in some ionic liquids [14]. One of important uranium(IV) species is $UO(OH)^+$, also formed under acid conditions.

Uranium(V) forms also a dioxo-cation [15]. In this oxidation state uranium disproportionate rapidly into its +6 and +4 state. The oxidation state +3 has no relevance under natural conditions. It is formed only in non-aquatic, strong reducing media.

Uranium(IV)

In solution, uranium(IV) shows a weak luminescence. This luminescence in the oxidation state +4 has been observed first by Kirishima et al. [16, 17]. The excitation is limited to a very small wavelength range around 245 nm. Depending on the complex formation of uranium(IV), the amount of excitation also varies.

As excitation source, a dye laser at 490 nm was used. To achieve the 245 nm, the output was frequency doubled. resulting in a short 245 nm laser pulse. An excimer laser at 308 nm was used to pump the dye laser. Due to the pulse duration of the excimer laser of about 20 ns, it was not possible to determine the decay time of the luminescence at room temperature. Therefore, the authors conclude that the luminescence lifetime of uranium(IV) may be shorter than 20 ns. It should be noticed here, that due to the short lifetimes of protactinium(IV) and americium(III), the determination of their excited states causes the same problems. The FWHM (full width at half maximum) of the excitation band was calculated to be 2.7 nm by use of measurements of the wavelength dependence of the excitation [18]. The lifetime increases to 149 ns in H_2O and to 198 ns in D_2O if the sample can be frozen to the temperature of liquid nitrogen [17].

Specially designed Nd:YAG pumped OPO systems may have pulse durations much shorter than 5 ns. In [18] the authors used a Nd:YAG laser system with a Pulse duration of about 8 ns. The OPO system, which was pumped by such laser pulses, showed a pulse duration of about 4 ns. Application of such a system to determine the uranium(IV) luminescence resulted in much shorter lifetimes [18] of 2.69 ± 0.08 ns.

At least 12 emission bands can be observed in the luminescence spectrum of uranium(IV). The most intense emissions are located at 319, 335, 410 and 525 nm. In addition to these emission peaks, lower intensities were located at 289, 292, 313, 321, 339, 346, 394 and 447 nm. It could be shown that the found emission maxima are in agreement with the possible transitions derived from the absorption spectrum [17].

In solid matrices, the luminescence of uranium(IV) was also studied. However, only one publication should be cited here. In this paper, the authors determined the luminescence of uranium(IV) in a $LiYF_4$ matrix [19]. For this matrix system in addition data of other actinides are available. The uranium(IV) excitation in the $LiYF_4$ matrix occurs at slightly higher energies and the maximum of the excitation was determined at a wavelength of 242 nm. In this matrix, the luminescence emissions were found to be at 262, 282, 304, 328, 334 nm. Also, two weak emission peaks can be observed at 430 and 490 nm. The lifetime of the luminescence in this sold matrix was determined to be about 17 ns. The lifetime does not depend on the temperature of the samples. The lifetime which was observed at 300 K and at 77 K was equal.

It seems the observation of uranium(IV) emissions in biosystems by use of single photon excitation at about 245 nm may be somewhat complicated. This should be connected to the strong UV-light absorption of biological samples. The intense energy application at wavelength of 245 nm and high laser pulse energies may also cause destructions in the sample material. The emission lifetime in aqueous media is very

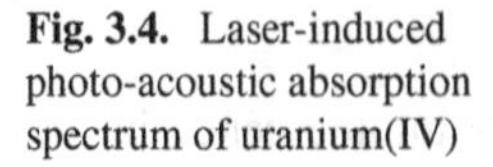

Fig. 3.4. Laser-induced photo-acoustic absorption spectrum of uranium(IV)

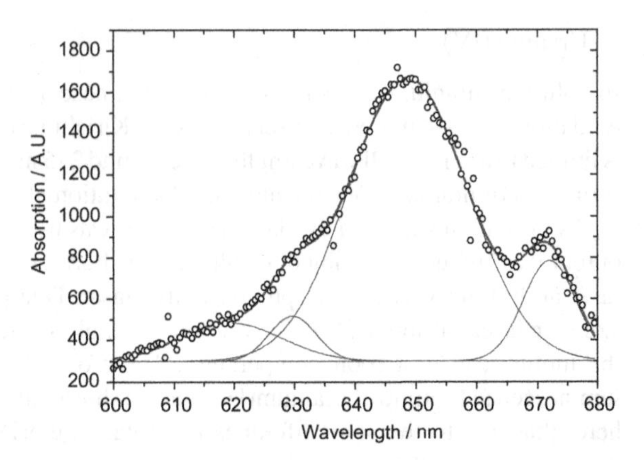

short and there is a need of special detection devices for the observation of short-lived emission spectra. In addition, quenching effects of the emission can be evident. This results in an additional signal and lifetime decrease. To observe the uranium(IV) luminescence in biological samples, make it clear in two photon excitation. However, this method requires a tuneable laser system, delivering ultrashort laser pulses in the femtosecond range.

Laser-induced photo-acoustic spectroscopy (LIPAS), a method of absorbance measurements, can provide more effort for detection of uranium(IV) species in biological systems. Using this method, excitation wavelength above 600 nm were applied and wavelength in the red and near infrared spectral range may cause less damage on the biological sample. Uranium(IV) species show about five higher time higher extinction coefficient than uranium(VI) species. For example a typical LIPAS spectrum of uranium(IV) is shown in Fig. 3.4. The uranium concentration in the sample was 1×10^{-4} M U^{4+} in 1 M $HClO_4$. This concentration seems to be high for biological samples. However, lower concentrations may be detectable with the development of noise reduction as well as by use of double beam methods.

Uranium(V)

The luminescence properties of uranium(V) have been described already [20]. The uranium(V) was prepared in a solution of 0.5M 2-propanol by use of an Hg-lamp. The spectrum was excited at 255 nm and the luminescence is emitted at around 440 nm with a decay time of 1.1 μs. As stated already for uranium(IV) it seems to be difficult to observe the emission of uranium(V) under such excitation conditions. Therefore, again absorption spectroscopic methods may be the method of choice. In literature several spectra are published. Here as example the spectra of uranium(V) in molten NaClx2CsCl is shown (Fig. 3.5) [21]. For the species $[UO_2(dbm)_2DMSO]^-$, $[UO_2(saloph)DMSO]^-$ and $[UO_2(CO_3)_3]^{5-}$ for example absorption maxima are given at 640 nm, 650 nm and 760 nm, respectively [22].

In case of uranium(V) an excitation wavelength of 255 nm is necessary. Under these conditions the same problem as for uranium(IV) luminescence detection in

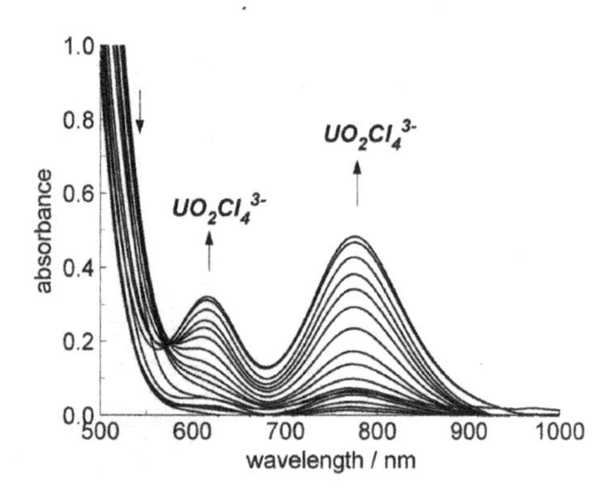

Fig. 3.5. Absorption spectra of formed uranium(V) species in molten NaClx2CsCl [21]. Reproduction with permission of DeGryter/IUPAC

biological samples exists. The detection of species by laser-induced emission spectroscopy seems to be not possible. As already stated, the use of two and three photon excitation may allow the detection of luminescence species of uranium(V) in biological systems in future.

Uranium(VI)

After excitation uranium(VI) emits luminescence from two emission levels. These levels are located at 21,270 cm^{-1} (470.1 nm) and 20,502 cm^{-1} (487.8 nm) [23]. The From the higher energy level a relatively low intensity of the emission is observed. Only about 5% of the total emission are emitted from this level. Therefore, emission from this level often cannot be detected. The explanation for this behaviour is seen in a so-called hot band. Energy transitions between two vibrational states occur and this results in an additional emission from a higher energy level. Such emission bands are strongly dependent on the temperature of the sample under study. If temperature of the sample decreases, also decreasing intensities of this band are observed. As the vibrational transition disappears, emissions from the higher level of the hot band are usually not detectable below about 120–150 K.

From the lower, normal energy level the emission reaches six vibration levels in the ground state. The energy distance between these levels is about 855 cm^{-1} [23]. Slight differences in this so-called band spacing can be observed, and these differences depend on the solution species under study and their solvation shell.

The FWHM (full width at half maximum) of the several transitions is not constant and values between 12 and 30 nm were found. In addition, the speciation of the uranium(VI) influences the FWHM. In case of the hydroxy species an increasing FWHM for higher hydrolyzed species like $UO_2(OH)_3^+$ and $UO_2(OH)_4$ is observed. Also, a broadening of the FWHM with increasing emission wavelength within the series occur.

In the literature no information about the intensity ratios of the bands from the emitting level 20,502 cm^{-1} are available. A first rough estimation on the intensity of the emission bands one should observe a decrease with increasing gap to the ground state vibrational levels, i.e. with increasing emission wavelength. However,

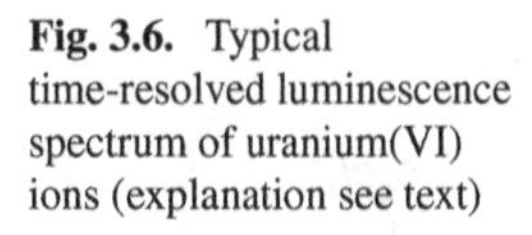

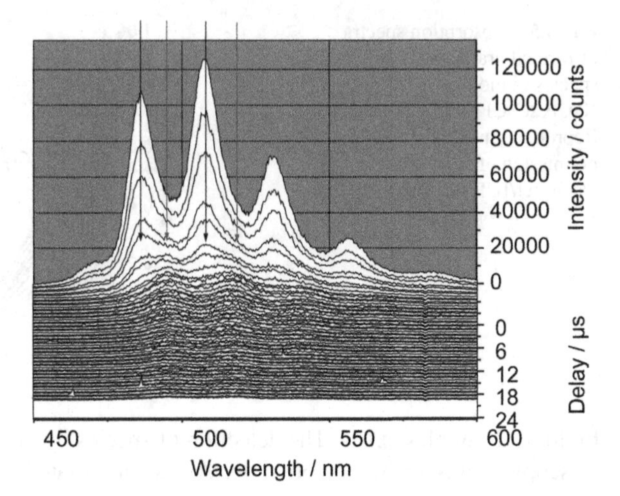

Fig. 3.6. Typical time-resolved luminescence spectrum of uranium(VI) ions (explanation see text)

often is observed, that the emission to the first vibrational level shows the highest luminescence intensity and mot the transition to the ground state.

In Fig. 3.6 a typical time resolved luminescence spectrum of uranyl ions is shown. An acquisition program is used which has allowed that three different, increasing delay steps (50, 100 and 500 ns) could be applied. A small influence of the exciting laser pulse of 266 nm could be detected at 532 nm, which corresponds to a second order image of the laser pulse. The solution contained 1×10^{-5} M UO_2^{2+}, The pH was adjusted to pH 4.5 and the ionic strength was set to 0.1M $NaClO_4$. The lifetime was fitted by used of a two exponential decay behavior and results in lifetimes values of 1.86 and 16.0 µs. This corresponds to the species UO_2^{2+} and UO_2OH^+ in expected the solution.

This shows that the uranium(VI) luminescence decay depends mainly on the speciation of the uranium(VI) ion. However, it should be noted that the decay time is also strongly influenced by the ionic strength of the medium [24]. In general, with increasing ionic strength an increase of the decay time can be observed. Therefore, the decay times summarized in Table 3.5 are compiled only for an ionic strength of 0.1 M.

Usually uranium(VI) species formed with organic carboxylic ligands show often no luminescence properties. In Table 3.5 some exceptions are listed. This may be connected to uranium binding not to the carboxylic group (i.e. phosphate, sulphate).

Uranium(VI) nitrate complexes are not listed in Table 3.5. However, they may by of great importance under environmental and reprocessing conditions. Due to the very weak nitrate complexation of uranium These complexes can be only determined in highly concentrated nitric acid solutions (>0.5 M HNO_3). Unfortunately, photochemical reactions of nitrate ions are well-known [45] and in addition it has been shown that nitrate ions also quench uranyl luminescence [46]. Therefore, no complete data set (emission and lifetime values) about the luminescence of the species $UO_2NO_3^+$ and $UO_2(NO_3)_2$ have been published [47].

Table 3.5 Emission wavelength and decay times of U(VI) solution species (lifetimes are usually compiled for room temperature, 25 °C)

Species	Emission wavelength (nm)						Decay time (ns)	References
UO_2^{2+}	472	488	510	535	560	587	1600	[25, 26]
	470	488	509	533	559	585	2000	[27, 28]
		489	510	535	560		1700	[29]
		488	509	534	560		900	[30]
	470	488	510	534	560	588	7900	[31][a]
$UO_2F_3^-/UO_2F_4^{2-}$		500	522	546	571		300,000	[32]
UO_2F^+		495	517	541			76,500	[33]
UO_2F_2		498	520	544			214,000	[33]
$UO_2IO_3^+$		494	515	538	565		<2000	[32]
$UO_2(IO_3)_2$		501	522	545	572		<2000	[32]
$UO_2IO_4^+$		503	524	547	574		<2000	[32]
UO_2OH^+	482	498	519	543	570	599	35,000	[25]
		497	519	544	570		80,000	[27]
		496	518	542	566		32,800	[29]
$UO_2(OH)_2$		488	508	534	558		<20,000	[27]
$(UO_2)_2(OH)_2^{2+}$	481	498	519	543	566	603	9500	[29]
		497	519	542	570	598	9000	[27]
	480	499	519	542	566		2900	[30]
$UO_2(OH)_3^-$	482	499	519	543	567	594	800	[27]
		506	524	555	568			[29]
$K_2UO_2(OH)_4$		491	510	531	552	579	154,000	[34][b]
$(UO_2)_3(OH)_5^+$	484	496	514	535	556	583	23,000	[27]
			515	536			6600	[29]
		500	516	533	554		7000	[30]
		498	514	534	557		19,800	[35]
$(UO_2)_3(OH)_7^-$		503	523	547	574		230,000	[27]
UO_2SO_4	477	493	515	538	**563**	590	4300	[36]
$UO_2(SO_4)_2^{2-}$		493	515	538			11,000	[36]
$UO_2(SO_4)_3^{4-}$		493	515	538			18,800	[36]
$UO_2H_2PO_4^+$		494	515	539	559		11,000	[37]
		493	514	538			11,100	[33]
UO_2HPO_4		497	519	543	570		6000	[37]
$UO_2PO_4^-$		499	520	544	571		24,000	[37]
$UO_2(H_2PO_4)(H_3PO_4)^+$		500	522	546	571		180,000	[32]
$UO_2(H_2PO_4)_2$		494	516	540			67,900	[33]

(continued)

Table 3.5 (continued)

Species	Emission wavelength (nm)						Decay time (ns)	References
$(UO_2)_x(PO_4)_y$	488	503	524	547	573	601	4700	[38]
$(UO_2)_x(K_yP_2O_7)_z$	484	499	520	544	568	598	75,000	[38]
UO_2HAsO_4		504	525	547			<1000	[26]
$UO_2H_2AsO_4^+$		478	494	514	539	563	12,200	[26]
$UO_2(H_2AsO_4)_2$		481	597	518	541	571	38,300	[26]
UO_2CO_3			520				35,000	[30]
$UO_2(CO_3)_3^{4-}$	466	485	505	525	548	573	30^1	
$Ca_2UO_2(CO_3)_3$	465	484	504	524			43	[39]
$UO_2OSi(OH)_3^+$		500	521	544	570		19,000	[40]
$UO_2C_3H_2O_4$	477	494	515	540	564	594	1240	[25]
$UO_2(C_3H_2O_4)_2^{2-}$	479	496	517	542	566	597	6480	[25]
UO_2ATP^{2-}	480	495	517	540	565	594	>20,000	[38]
$UO_2(C_6H_{11}O_6PO_3)$	479	497	519	543	569	598	130	[41]
$UO_2(HGly)^{2+}$	476	492	513	537	562	587	3350	[42]
$UO_2(HGly)_2^{2+}$	479	495	517	541	565	594	690	[42]
$UO_2(Gly)_4^{2+}$		495	516	538	561		–	[43]$^+$
UO_2HTr^{2+}	475	492	514	539	565	591	810	[44]
$UO_2H_2Tr_2^{2+}$	476	494	516	541	566	595	330	[44]
$UO_2H_3PTr^{2+}$	478	494	515	540	564	594	17,400	[44]
$UO_2H_2PTr^+$	479	496	517	541	566	595	4900	[44]
UO_2HPTr	484	502	523	547	573	601	540	[44]

Gly = glycine; Tr = L-threonine; PTr = O-phospho-L-threonine
[a] 1 M $HClO_4$; values are rounded, the errors in emission wavelength are 2; 0.8; 0.6; 0.6; 0.6 and 2 nm, the error of the decay time is 0.7 μs
[b] Measured at 150 K
[+] 0.019 M solution, excitation at 416.5 nm
[1] Measured at 278.1 K

The luminescence of uranium(VI) may be influenced by many effects in the solution. A large number of quenching substances have been studied in the past. The solvent water as well as carbonate ions, that play an important role in environmental samples, should be pointed out firstly. Iron and chloride ions are often present in solution and show dynamic quench properties in relation to the luminescence of uranium(VI). A large list of known quenchers and associated quenching constants are summarized in [48]. The phenomenon of quenching processes can be used in order to assay the concentration of the quenching substances.

It should be noted, that quench effects are distinguished into two phenomena—static and dynamic quench effects. Static quench effects are caused by complex formation phenomena, where the formed complex does not emit any luminescence, and therefore no change in the lifetime is observed. Dynamic quench effects are caused by interactions between the excited emitter and the quencher, where energy is transferred to the quencher. This leads to a decrease of the intensity of the emitted light as well as the lifetime. By measuring luminescence intensity as well as lifetime one can distinguish between both effects and it is also possible to determine stability constants in systems where dynamic quench effects occur. In addition, dynamic quench effects are temperature dependent.

Table 3.5 shows a wide variety in luminescence emission bands as well as in decay times of uranium(VI) depending on the solution species. On the other side this can be used for determination of the uranium species formed. Nevertheless, the table also shows, that the observed luminescence decay times may be influenced by a number of factors, leading to differences in the published data. The Round Robin Test [31] has shown that there is a need in more comparative measurements to get consistent data sets in the luminescence determination of uranium(VI) species. There are three important conclusions of this test. First, the determined wavelength of the uranium species used for this test are in good agreement between the several participants. Second, the lifetimes scatter in a relatively large range. This was explained with a dependence on the time between sample preparation and measurement. Not all participants could measure their samples with the same time delay. In addition to this some participants could not determine minor species. An explanation was, that species could adsorb on the surface of the used cuvettes.

References

1. Lieser KH (1991) Einführung in die Kernchemie. VCH Weinheim, 3rd edn., 107–109
2. Rössler HJ (1991) Lehrbuch der Mineralogie, 5th edn. Deutscher Verlag für Grundstoffindustrie, Leipzig, 184
3. Lederer DC et al (1968) Table of isotopes, 6th edn. Wiley, New York
4. Frondel C (1958) Systematic Mineralogy of uranium and thorium, Geological Survey Bulletin 1064. US Government Printing Office, Washington
5. Enghag P (2004) Encyclopedia of the elements. Wiley-VCH Verlag GmbH & Co., Weinheim, p 1195
6. Evans RD, Goodman C (1941) Radioactivity of rocks. Geol Soc Am Bull 52:459–490
7. Spalding RF, Sackett WM (1972) Uranium in runoff from gulf of Mexico distributive province—anomalous concentrations. Science 178:629–631
8. Hoffmann J (1943) The bio-element uranium in plant and animal kingdoms as well as in human organisms. Biochem Z 313:377–387
9. Stewart DC, Bentley WC (1954) Analysis of uranium in sea water. Science 120:50–51
10. Sparovek RBM, Fleckenstein J, Schnug E (2001) Issues of uranium and radioactivity in natural mineral waters. Landbauforschung Völkenrode 51:149–157
11. U.S. Environmental Protection Agency (1991) Review of Relative Source contribution analysis, report prepared by Wade Miller Associates, Inc. for the Office of drinking water
12. Fisenne IM, Perry PM, Decker KM, Keller HW (1987) The daily intake of U-234, U-235, U-238, Th-228, Th-230, Th-232, and Ra-226, Ra-228 by New-York-City residents. Health Phys 53:357–363
13. Opel K, Weiß S, Hübener S, Zänker H, Bernhard G (2007) Study of the solubility of amorphous and crystalline uranium dioxide by combined spectroscopic methods. Radiochim Acta 95:143–149

14. Trubert D, Le Naour C, Jaussaud C (2002) Hydrolysis of protactinium(V). I. Equilibrium constants at 25 degrees C: a solvent extraction study with TTA in the aqueous system $Pa(V)/H_2O/H^+/Na^+/ClO_4^-$. J Sol Chem 31:261–277
15. Nikitenko SI, Cannes C, Le Naour C, Moisy P, Trubert D (2005) Spectroscopic and electrochemical studies of U(IV)-hexachloro complexes in hydrophobic room-temperature ionic liquids [BuMelm][Tf_2N] and [$MeBU_3N$][Tf_2N]. Inorg Chem 44:9497–9505
16. Kirishima A, Rimura T, Tochiyama O, Yoshida Z (2003) Luminescence study of tetravalent uranium in aqueous solution. Chem Commun 910–911
17. Kirishima A, Kimura T, Nagaishi R, Tochiyama O (2004) Luminescence properties of tetravalent uranium in aqueous solution. Radiochim Acta 92:705–710
18. Geipel G (2004) Fluorescence properties of uranium(IV). In: Bernhard G (ed), Annual report 2003, Institute of Radiochemistry, FZR- 400, 1
19. Godbole SV, Page AG, Sangeeta A, Sabharwal SC, Gesland JY, Sastry MD (2001) UV luminescence of U^{4+} ions in $LiYF_4$ single crystal: observation of 5f(1)6d(1) -> 5f(2) transitions. J Lumin 93:213–221
20. Steudtner R, Arnold T, Großmann K, Geipel G, Brendler V (2006) Luminescence spectrum of uranyl(V) in 2-propanol perchlorate solution. Inorg Chemi Commun 9:939–941
21. Volkovich VA, Aleksandrov DE, Griffiths TR, Vasin BD, Khabibullin TK, Maltsev DS (2010) On the formation of uranium(V) species in alkali chloride melts. Pure Appl Chem 82:1701–1717. https://www.iupac.org/publications/pac/pdf/2010/pdf/8208x1701.pdf
22. Mizuoka K, Tsushima S, Hasegawa M, Hoshi T, Ikeda Y (2005) Electronic spectra of pure uranyl(V) complexes: characteristic absorption bands due to a (UO_2^+)-O-V core in visible and near-infrared regions. Inorg Chem 44:6211–6218
23. Bell JT, Biggers RE (1968) Absorption spectrum of uranyl ion in perchlorate media. 3. Resolution of ultraviolet band structure—some conclusions concerning excited state of UO_2^{2+}. J Mol Spect 25:312–329
24. Billard I, Rustenholtz A, Semon L, Lützenkirchen K (2001) Fluorescence of UO_2^{2+} in a non-complexing medium: $HClO_4/NaClO_4$ up to 10 M. Chem Phys 270:345–354
25. Brachmann A, Geipel G, Bernhard G, Nitsche H (2002) Study of uranyl(VI) malonate complexation by time resolved laser-induced fluorescence spectroscopy (TRLFS). Radiochim Acta 90:147–149
26. Rutsch M, Geipel G, Brendler V, Bernhard G, Nitsche H (1999) Interaction of uranium(VI) with arsenate(V) in aqueous solution studied by time-resolved laser-induced fluorescence spectroscopy (TRLFS). Radiochim Acta 86:135–141
27. Laszak I, Moulin V, Chr. Moulin P, Mauchien P (1997) 1. Technical report, EC project Effects of humic substances on the migration of radionuclides. CEA contribution to Task 2
28. Moulin Chr., Laszak I, Moulin V, Tondre C (1998) Time-resolved laser-induced fluorescence as a unique tool for low-level uranium speciation. Appl Spect 52:528–535
29. Eliet V, Bidoglio G, Omenetto N, Parma L, Grenthe IJ (1995) Charactization of hydroxide complexes of uranium(VI) by time-resolved fluorescence spectroscopy. Chem Soc Faraday Trans 91:2275–2285
30. Kato Y, Meinrath G, Kimura T, Yoshida Z (1994) A study of U(VI) hydrolysis and carbonate complexation by time-resolved laser-induced fluorescence spectroscopy (TRLFS). Radiochim Acta 64:107–111
31. Billard I, Ansoborlo E, Apperson K, Arpigny S, Azenha ME, Birch D, Bros P, Burrows HD, Choppin G, Coustin L, Dubois V, Fanghänel T, Geipel G, Hubert S, Kim JI, Kimura T, Klenze R, Kronenberg A, Kumke M, Lagarde G, Lamarque G, Lis S, Madic C, Meinrath G, Moulin C, Nagaishi R, Parker D, Planque G, Scherbaum F, Simoni E, Sinkov S, Viallesoubranne C (2003) Aqueous solutions of uranium(VI) as studied by time-resolved emission spectroscopy: a round-robin test. Appl Spectrosc 57:1027–1038
32. Karbowiak M, Hubert S, Fourest B, Moulin C (2004) Complex formation of uranium(VI) in periodate solutions. Radiochim Acta 92:489–494
33. Kirishima A, Kimura T, Tochiyama O, Yoshida Z (2004) Speciation study on complex formation of uranium(VI) with phosphate and fluoride at high temperatures and pressures by time-resolved laser-induced fluorescence spectroscopy. Radiochim Acta 92:889–896

34. Tits J, Geipel G, Steudtner R, Eilzer M (2007) TRLFS measurements of U(VI) sorbed on CSH phases under alkaline condition. In: Bernhard G (ed) Annual report 2006. Institute of Radiochemistry; FZD-459, 55
35. Sachs S, Brendler V, Geipel G (2006) Uranium(VI) complexation by humic acid under neutral pH conditions studied by laser-induced fluorescence spectroscopy. Radiochim Acta 95:103–110
36. Geipel G, Brachmann A, Brendler V, Bernhard G, Nitsche H (1996) Uranium(VI) sulfate complexation studied by time-resolved laser-induced fluorescence spectroscopy (TRLFS). Radiochim Acta 75:199–204
37. Scapolan S, Ansoborlo E, Moulin C, Madic C (1998) Investigations by time-resolved laser-induced fluorescence and capillary electrophoresis of the uranyl-phosphate species: application to blood serum. J Alloys Comp 271–273:106–111
38. Geipel G, Bernhard G, Brendler V, Reich T (2000) 5th International Conference on Nuclear and Radiochemistry. Extended Abstracts, vol 2, Pontresina, Switzerland, 473–476
39. Bernhard G, Geipel G, Reich T, Brendler V, Amayri S, Nitsche H (2001) Uranyl(VI) carbonate complex formation: validation of the $Ca_2UO_2(CO_3)_{(3)}$(aq.) species. Radiochim Acta 89:511–518
40. Moll H, Geipel G, Brendler V, Bernhard G, Nitsche H (1998) Interaction of uranium(VI) with silicic acid in aqueous solutions studied by time-resolved laser-induced fluorescence spectroscopy (TRLFS). J Alloys Comp 271–273:765–769
41. Koban A, Geipel G, Roßberg A, Bernhard G (2004) Uranium(VI) complexes with sugar phosphates in aqueous solution. Radiochim Acta 92:903–908
42. Günther A, Geipel G, Bernhard G (2007) Complex formation of uranium(VI) with the amino acids L-glycine and L-cysteine: a fluorescence emission and UV-Vis absorption study. Polyhedron 26:59–65
43. Alcock ND, Flanders DJ, Kemp TJ, Shand MA (1985) Glycine complexation with uranyl-ion—absorptiometric, luminescence, and X-ray structural studies of tetrakis(glycine)dioxouranium(VI) nitrate. J Chem Soc Dalton Trans 517–521
44. Guenther A, Geipel G, Bernhard G (2006) Complex formation of U(VI) with the amino acid L-threonine and the corresponding phosphate ester O-phospho-L-threonine. Radiochim Acta 94:845–851
45. Mack J, Bolton JR (1999) Photochemistry of nitrite and nitrate in aqueous solution: a review. J Photochem Photobiol A 128:1–3
46. Deniau P, Decambox P, Mauchien P, Moulin C (1993) Time-resolved laser-induced spectrofluorometry of UO_2^{2+} in nitric-acid solutions –preliminary results for online uranium monitoring applications. Radiochim Acta 61:23–28
47. Couston L, Pouyat D, Moulin Chr., Decambox P (1995) Speciation of uranyl species in nitric-acid medium by time-resolved laser-induced fluorescence. Appl Spectrosc 49:349–353
48. Hoffman MZ, Bolletta F, Moggi L, Hug GL (1989) Rate constants for the quenching of excited-states of metals complexe in fluid solution. J Phys Chem Ref Data 18:219–544

Chapter 4
Uranium and Relevant Bioligands

To get more deep insights in the complex coordination chemistry of uranium with biological compounds like proteins, model complex formations should be characterized. As example, how such studies can be performed, the investigation of interactions between aspartyl-rich pentapeptides and tetravalent actinides [1] should be cited. However, this work does not include uranium. From results for tetravalent neptunium one can conclude the direction of uranium(IV) results.

Mitochondria produce adenosine triphosphate (ATP). The energy-transfer functions of this compound interact mostly intracellular. Binding of uranium and neptunium to ATP has been reported in [2, 3]. It was found by TRLFS studies that complex formation is accompanied by static quenching processes as well as strong dynamic quenching processes. These processes decrease the fluorescence intensity. Due to the fact that only dynamic quench processes influence the luminescence lifetime, measurements of the lifetime data can be used to calculate the effect of pure static quenching and then to estimate thermodynamic values. The derived stability constants depend strongly on pH of the solution. Therefore, it was concluded that protons are involved in the chemical equilibrium. The formation constant was assigned to be $\log K_A = -3.80 \pm 0.44$ for the 1:1 complex.

Antibodies are often used to study molecular interactions due to their properties in specific targeting. It has been published about the synthesis of 5-isothiocyanato-1,10-phenanthroline-2,9-dicarboxylic acid (DCP). This compound forms complexes with uranyl ions [4]. It has been also reported that the formed complex coupled to a carrier protein interacts with the polyclonal antibodies 8A11; 10A3 and 12F6. The dissociation constants for the respective UO_2(DCP)-antibody complexes were calculated to be 5.5, 2.4 and 0.9 nM. The bond strength of DCP towards antibodies without any coupled metal is about three orders of magnitude lower. A very sensitive detection of UO_2(DCP) is estimated for these antibodies if compared to binding constants with other metal-DCP complexes. The conclusion of this study was that uranium contamination could be monitored and controlled by use of specific uranium-binding capabilities. This strategy may support concepts in developing immunoassays for UO_2^{2+}.

G. Geipel, *Uranium and Plant Metabolism*, SpringerBriefs in Biometals,
https://doi.org/10.1007/978-3-030-80815-0_4

References

1. Jeanson A, Berthon C, Coantic S et al (2009) The role of aspartyl-rich pentapeptides in comparative complexation of actinide(IV) and iron(III). Part 1. New J Chem 33:976–985
2. Geipel G, Bernhard G, Brendler V, Reich T (2000) 5th International Conference on Nuclear and Radiochemistry. Extended Abstracts, vol 2, Pontresina, Switzerland, 473–476
3. Rizkalla EN, Netoux F, Dabosseignon S, Pages M (1993) Complexation of neptunium(V) by adenosine phosphates. J Inorg Biochem 51:701–703
4. Blake RC, Pavlov AR, Khosraviani M, Ensley HE, Kiefer GE, Yu H, Li X, Blake DA (2004) Novel monoclonal antibodies with specificity for chelated uranium(VI): isolation and binding properties. Bioconjugate Chem 15:1125–1136

Chapter 5
Uranium Uptake, Storage and Metabolism by Plants

One of the most important questions for uptake of heavy metals by plants is their bioavailability. Ernst [1] has reported on this from the viewpoint of decontamination. However, the basic questions are also valid for uptake of heavy metals:

- bioavailability in the soil
- possibilities and limits of plants for phytoremediation
- vegetation and re-cultivation of contaminated soils

From a review of data for transfer of a radionuclide of heavy element Ehlken [2] concludes that transfer factor not only depends on the total concentration of the element under investigation. Other factors influence the uptake and therefore the transfer factor.

Plant studies towards uranium often deal with two topics, one is the metal uptake by transfer factors the other one is the use of plants as bio-monitors. Uranium uptake and influence of this metal on living matter has been reviewed already [3]. However, speciation of uranium has not been studied in detail. Sequential extraction techniques have been used. These methods allow only a very sketchy information about the involved species.

Last not least, it should be mentioned, that in soils may occur an upward transport of radionuclides. This was reported by Perez-Sanchez [4]. This upward transport may vary seasonally. However, it cannot be stated that only a downward transport occurs connected to the water flux to the groundwater.

Under natural conditions hyphae and roots are the first organs of a plant which have contact to the uranium in the environment. Therefore, is the study of their contribution to the uranium uptake of immanent importance. Rufyikiri and Co-workers report on investigations about uranium uptake by carrot roots [5]. Tracer amounts of the isotope ^{233}U were used as spike. For the hyphae they found a 5.5 and 9.6 higher uptake than for mycorrhizal and non-mycorrhizal roots. Also, it was found that the uranium flux rate is higher in the fungal hyphae. In contrast to this, intra-radical hyphae contribute significantly to uranium immobilization.

G. Geipel, *Uranium and Plant Metabolism*, SpringerBriefs in Biometals,
https://doi.org/10.1007/978-3-030-80815-0_5

Vandenhove and coworkers [6] tried to establish a correlation between soil parameters and uranium soil solution concentration. It was found that only at pH > 6 a linear correlation between pH and uranium concentration exits and that there exist a correlation between the solid–liquid distribution coefficient and the content of organic matter as well as amorphous iron. In a second part of this study [7] the uptake of uranium by ryegrass growing on these soils was studied. Transfer factors between 0.0003 and 0.034 kg kg^{-1} were found. A relation between uranium concentration and uptake was not found. However, uranium speciation influences the uptake. Preferred species are UO_2^{2+}; $UO_2PO_4^-$ and uranyl carbonates, respectively.

Soils and sediments from a uranium processing facility showed U concentrations with an average of 630 mg U kg^{-1}. Highest uranium accumulation was found in aquatic mosses with concentrations up to 12,500 mg U kg^{-1} dry weight (DW). This is about 1% of the dry mass. Several macrophytes (*Phragmites communis, Scripus fontinalis* and *Sagittaria latifolia*) accumulate uranium also very good. Usually plant roots contain higher amount of uranium than shoots and leaves. The roots of *Impatiens capensis* accumulated up to 1030 mg kg^{-1}, followed by the roots of *Cyperus esculentus* and *Solidago speciosa.* Transfer factors were determined for roots of *P. communis* 17.4, *I. capensis* 3.1 and for the whole plant *S. fontinalis* 1.4. Uranium in plant tissue is often correlated to strontium This element is chemically and physically similar to calcium (Ca) and magnesium (Mg), which were also positively-correlated with U. The correlation between U and these plant nutrient elements, including iron (Fe), suggest an active uptake mechanism for uranium accumulation [8].

Arabidopsis haleri growed on a former U mining site and were harvested during a field trip. The uranium accumulation and tolerance of plant samples was compared by Viehweger [9] by laboratory trials versus plants growing on the original site. By sequential extraction, the solubility of several metals was determined. Low bioavailability of the micronutrient iron correlates with the uptake of the non-essential uranium. Iron occurs in nature mostly as Fe(III) oxides and hydroxides. Therefore, a requirement for the uptake is the reduction of Fe(III) to Fe(II) at the root surface. Reduction of uranium can accompany this process. A similar reduction of UO_2^{2+} to U^{4+} under laboratory conditions could be shown by photoacoustic measurements. The accumulation of uranium by *Arabidopsis* plants growing on the mining site was found to be 35 mg U kg^{-1} (dry weight) in roots and 17 mg U kg^{-1} (dry weight) in shoots. With regard to the bioavailability of uranium in the natural environment, the soil-to-plant transfer factor (TF) was calculated to about 1.2 for roots and 0.6 for shoots, respectively. Under laboratory conditions plants were grown hydroponically. The uranium content was found to be 100-fold higher in roots and tenfold higher in shoots. An iron deficiency of hydroponically grown plants was recognized as reason for this strong increase of uranium uptake.

Root elongation measurements were used to get more information about mechanisms of U to the tolerance index (TI). Additionally, the chlorophyll content was determined in order to have some impacts on basic photosynthetic traits. In native grown plants a chlorophyll a/b ratio around 7 was found. In contrast the ratio of hydroponically grown plants dropped down during the growth cycle. This was caused by the above-mentioned lack of iron. From chlorophyll extracts from U containing leaves were taken fluorescence spectra. An additional peak was observed which was assigned to a flavonoid. This can be seen as an indication that deficiency of essential metal ions can enhance the uptake of non-essential metal ions.

Uranium may cause oxidative stress in plants. Aranjuelo [10] studied the effect of 50 μM uranyl on Photosystem II (PSII) and parameters connected to water transport especially leaf transpiration and aquaporin gene expression of *Arabidopsis* wild type (WT) and mutant plants. Uranium exposure induced photosynthetic inhibition. This causes an electron sink/source imbalance which is connected to PSII photoinhibition in the mutants. Only the wild type did not show any damage in PSII when uranium was present. A relationship between leaf U content and leaf transpiration was found, supporting the relevance of water transport in heavy metals partitioning and accumulation in leaves, with the consequent implication of susceptibility to oxidative stress.

Haas [11] has studied the sorption of uranium in lichen from aqueous media. In the pH range 4–5 were observed the highest uptake values. He performed an analysis of micro-areas. This has shown that the uranium distribution is heterogenous. High local concentrations in the upper cortex were observed. It was also stated that the uranium concentration seems to be correlated to the phosphor concentration. The authors conclude that biomass-derived phosphate ligands or functional groups on the surface may be involved.

Markich [12] published an overview on speciation and bioavailability of uranium in aquatic systems. It was suggested from the data pool that the free uranyl ion UO_2^{2+} and its first hydrolysis product UO_2OH^+ should be uranium species that control the bioavailability mainly. Complexes formed with inorganic or organic ligands, like phosphate ions, carbonate ions and humic substances were addressed to be less important. The plant/soil concentration ratios for uranium and thorium were summarized by Sheppard [13]. A wide variability of the ratios was reported. Causes for this were seen mainly in the soil type and also and in different plants used in the studies. In general, it can be stated that means are of 0.0045 for uranium and 0.0036 for thorium. In addition, it seems that the data support the suggestion of a relationship between the uranium and the phosphorous concentration in the fresh sample. However, a correlation coefficient of about 0.75 was found, which indicates a correlation to be not very strong. The following basic information about the concentration ratios were seen to be important:

1. Coarse textured mineral soils have higher concentration ratios than fine textured soils.
2. Root crops show higher values than cereal grain crops.
3. Increasing contaminant concentration decreases the concentration ratios.

Time-resolved laser-induced fluorescence spectroscopy (TRLFS) was suggested to be one method to determine the speciation of uranium(VI). However, this viewpoint may be not complete. Uranium forms strong complexes with phosphate and the solubility of uranyl phosphate species is rather limited. Also, phosphate is an important constituent of the nutrition media. Secondly plants are able to exudate agents to make insoluble components bioavailable. As organic acids form soluble complexes with actinides, the uptake by plants may increase if such species are released from roots.

Uptake by tomato plants growing in uranium-contaminated soil has been examined [14]. The soil was contaminated artificially with several concentrations of uranium. Due to a lack of exact experimental details, only two qualitative conclusions may be drawn:

1. The concentration of uranium decreases in the order root > stem > leaf. This finding would be expected for all plants as the roots have direct contact to the bioavailable uranium.
2. The uranium concentration in the roots is relatively high compared to that in the r contaminated soil. This observation means that tomato plants are "uranium accumulators".

Conclusion (1) was, however, not confirmed in another study [15].

A wide range from 7.2×10^{-5} to 1×10^{-3} has been found for soil-to-plant transfer factors for uranium [16]. Several years ago, transfer factors for uranium into vegetables were reported and the values are somewhat higher than the above-mentioned data [17].

As actinides show different oxidation states also different uptake levels should be expected. The uptake of elements belonging to the uranium decay series was studied on plants growing on a mine mill [18]. The highest uptake for uranium followed by thorium was found for plants growing on weathering tailings. The authors used mixed groups of plants. Therefore, probably the data are not comparable to others. The behavior of ^{238}U-series radionuclides in plants and soils was published recently by Mitchell and co-workers [19] in a comprehensive documentation.

To determine uranium levels in plants from a milling site at Köprübasi in Turkey the authors applied neutron activation analysis [20]. The uranium level in all plants except wheat was found to be above 0.6 ppm. In the ash of plants, which were collected at a U-deposit, an enrichment of uranium up to 16 ppm was found.

If plants show an uptake of radionuclides the formed species inside the plant are different than those which are present in soil [21]. It was also concluded in this review paper that uptake generally depends on the concentration of radionuclides in the soil. If plants collect more uranium than present in the natural background then mechanisms of accumulation are assumed [22]. An example for such plats is black spruce. Uranium hyperaccumulators with a shoot/root ratio < 1 were not identified up to now [23].

Soil containing acetic, malic and citric acid increased uptake of uranium by plants was observed [24]. If soil contains for example citric acid then it was found that the uranium concentration of uranium in the shoots of *Brassica juncea* and *B. chinensis* increases dramatically by a factor of about one thousand. Concentrations of 5 mg/kg were detected under non-influenced conditions and under influence of citric acid more than 5000 mg/kg. An explanation of this extremely high uptake was seen by the biodegradation of citric acid. A possible degradation product should be propionic acid [24]. However, in other cases the nature of the degradation products was mostly unknown [25, 26].

With respect to relationships between metal speciation and plant response there is only little known up to now. The first study including the speciation of uranium during the growth of plants was carried out in 1998 by Ebbs and coworkers [27]. Under several conditions the behavior of pea (*Pisum sativum*) was studied. The speciation of uranium in the hydroponic solution at the beginning of experiments was calculated with the program GEOCHEM-PC. A modified Johnson' nutrient solution was used. Plants with a normal solution with phosphate ions but without uranium were pretreated for 10 days. After this start period the plants were transferred to nutrient solutions containing no phosphate ions, but uranyl ions were added. The plants were grown for additional seven days before harvesting. Roots and shoots were rinsed and dried and afterwards digested with nitric acid. It was found that the uranium uptake showed an influence of the pH value. At pH 5, when uranium mainly was present as the free uranyl cation (UO_2^{2+}), the root concentration was found to be higher than that in the shoots. However, it seems according to recent speciation calculations, that also hydroxy species could be formed. Other plant species, *Phaseolus acutifolius* and *Beta vulgaris* showed the highest uranium uptake under these conditions. Information on the speciation in these plants is not available due to lack in detection methods at this time.

A later study of uranium speciation in plants was performed with lupines (*Lupinus angustifolius*) [28]. Experiments have been realized in two series. In one series the plants were grown in soil. The other series used hydroponic solution. The speciation in the two growing compartments was found to be different: In the hydroponic solution are uranyl-hydroxy species the majority. In the soil pore water carbonate species dominate due to the stronger contact to the atmosphere and to containing carbonate minerals. However, speciation in several parts of the plant was stated to be the same. This lead to the conclusion that the speciation in the plant parts is independent of the species present in the original solution. It was found that uranium is bound mainly to phosphoryl groups in plant compartments. This could be confirmed by EXAFS measurements and TRLFS studies.

Laurette and co-workers gained recently strong progress in knowledge of uranium speciation in plants [29]. Plant species of oilseed rape, *Brassica napus*, sunflower, *Helianthus annuus* and wheat, *Triticum aestivum* were contacted with different uranyl species. Variation of pH values led to different amount of uranium accumulation in plant roots and the leaves showed differences in translocation behavior. In scenario 1, which consisted mainly in UO_2^{2+} and carbonate species at pH 4, the uptake resulted in uranium precipitation on root epidermis. Speciation calculations for uranium under these pH conditions show only formation of UO_2CO_3 and hydroxy species as complexed species. Small amounts of uranium were determined in the soluble fraction and could be transported via apoplasm or symplasm. In scenario 2 phosphate ions were present in the nutrition solution. This led to a moderate to low uranium uptake and translocation of uranium was negligible. Under these conditions already precipitation of uranium may occur. Carbonate or citrate containing media at neutral or acid pH were used in scenario 3. In contrast to the findings in the other media the uptake to the roots and translocation to the shoots were extremely increased. At the moment the speciation of uranium existing in different plant tissues could not be determined. The detection of possible uranyl binding proteins in soluble, cellular fractions separated by sequential extraction followed by gel chromatography seems to be an interesting way to increase the knowledge about transport processes in plants. Next steps should be the identification of these proteins and to become insights in their possible functions. Rossberg et al. reported that uranium(VI) is able to bind in DNA to two phosphoryl groups [30]. The surprising result was that these two phosphoryl groups are located in the two strands of the DNA helix. This means uranium(VI) connects the two strands. Cell division will be not possible due to the combined strands and uranium release requires DNA disruption. This may have fundamental impact on the life of the cell. Combined sensitive metal detection via inductively coupled plasma mass spectrometry (ICP-MS) with MS based proteomics methods is a new approach is increase knowledge in metallomics and metalloproteomics science. The possibilities and power of new setups using high-throughput tandem mass spectrometry (HT-MS/MS) and ICP-MS to characterize cytoplasmic metalloproteins from exemplary microorganisms and the identification of uranyl binding proteins was recently and impressively demonstrated Cvetkovic and co-workers [31]. Relatively new reviews discussing this advanced field of metal uptake and speciation were reported by Maret [32] and Yannone and co-workers [33].

Knowledge in metal speciation in plant tissues and cells, working on a molecular level is extremely important in future. Plant cell cultures grown in a laboratory are applicable models for characterization of metal chelating metabolites, cellular transportation as well as signaling or accumulation.

Suspension cultures of canola (*Brassica napus*) were used by Viehweger [34]. Uranium influence on the cellular glutathione pool in reduced or oxidized form (GSH, reduced form; GS-SG oxidized form) were studied. The consequences for the redox status in cell tissues were discussed. After cellular uptake, uranium(VI) probably can form cyclic complexes with glutathione via the carboxylic acid groups of GSH [35] whereas uranium(IV) is should be coordinated at the sulfhydryl group. The latter already was postulated for cysteine [36]. Initially formed uranium(V) species disproportionate into uranium(VI) and uranium(IV). This may explain that only 37% (relative to the initial uranium concentration) reduced uranium(IV) was found in the cells. After 2 h of contact the highest uranium(IV) level was detected. Within 10 h the uranium(IV) fraction decreases again. Applying the Nernst equation the calculated redox potential in the solution for lower uranium concentrations (10 μM) showed a shift towards more oxidizing conditions. However, the opposite effect was detected at higher concentrations (50 μM). This demonstrates eminently that the redox behaviour of this metal plays an important role in cellular uranium speciation and coordination.

The environmental speciation also should be taken into account. An overview was given by Choppin [37]. Different redox states of uranium were detected at the roots of a naturally adapted variety of *Arabidopsis halleri* [8]. This plant was collected on a former uranium site in Johanngeorgenstadt (Germany), which was not yet revitalized. A correlation between low bioavailability of the nutrient iron and the uptake of non-essential uranium was found by use of sequential extraction of soil samples collected together with the plant material. Reduction of Fe^{3+} to Fe^{2+} on the roots is necessary for iron uptake by plants and uranium reduction could take place in concurrence to iron reduction. Reduction of uranium(VI) to uranium(IV) was confirmed by Laser-induced photo-acoustic spectroscopy (LIPAS). Reduction of uranium(VI) to uranium(V) is proposed to occur on the roots. In a following disproportionation step uranium(IV) on the roots is generated.

On rock piles *Arabidopsis* plants grow under natural conditions. These plants are able to accumulate ~35 mg/kg uranium (dry weight) in the roots and ~17 mg/kg in the shoots [8]. The bioavailability of uranium in the soil has been considered and the soil to plant transfer factor was then estimated to be 1.2 for the roots and 0.6 for the shoots. Uranium uptake was about 100-fold higher for roots and tenfold higher in shoots if plants were grown in the laboratory under hydroponic conditions and when the concentrations are comparable to natively grown plants. The authors reasoned, that this radical increase in uranium uptake should be a consequence of the iron deficiency in the hydroponic system. Fluorescence spectra were taken from leaves and chlorophyll extracts. These spectra indicated the occurrence of a flavonoid after uranium contact. It is known, that flavonoids can act as a possible chelator of uranium.

The comparison with cellular behaviour of other heavy metals should also a possible procedure.

Iron is probably a good example due to interactions between the uptake of the micronutrient iron and uranium, as already stated [8]. Caldwell and co-workers [9] published also a significant relationship between iron and uranium plant concentrations (correlation: 0.783; r-squared: 0.613). However, the correlation coefficients are not too high. Some metals ions act as micronutrients and therefore, their bioavailability should also have an impact on plant uranium uptake. The knowledge concerning transition metal chelation and translocation inside plants and plant cells has been raised enormously during the last years. Kraemer et al. [38] or Kobayashi [39] published important reviews. Chelators like citrate, nicotinamine or phenolics were discussed. These ligands are able to chelate uranium. Most of the transport mechanisms are shared by various metals and so uranium may also be included. In Fig. 5.1 different interactions between the uranium, soil and plant are shown. All these interactions influence the metal speciation. However, this occur in a sophisticated network.

An introduction about mechanism of interaction and uptake of heavy metals by plants is given in the review by Viehweger [41]. Unfortunately, this review does not include uranium. However, it may be used as reference booklet, which have to be included in such studies besides metal tolerance and toxicity.

Doustaly [42] used J-Chess modeling for the prediction of uranium speciation and exposure conditions which may affect the uranium bioavailability for plants. To confirm this model Arabidopsis thaliana plants were exposed to uranium under hydroponic conditions. Root response was characterized with complete *Arabidopsis* transcriptome microarrays (CATMA). At three different times after exposure the expression of 111 genes was studied. The associated biological processes were examined by real-time quantitative RT-PCR. Uranium exposure has been found to

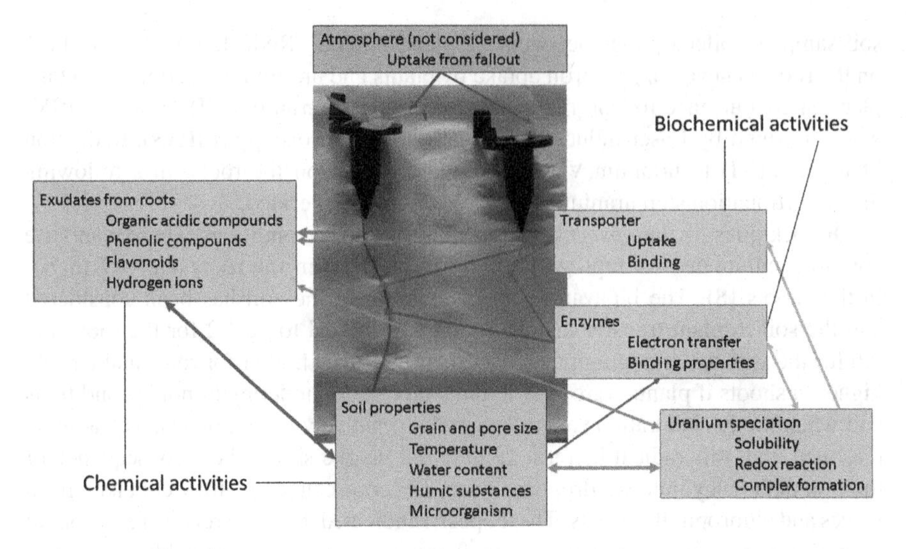

Fig. 5.1 Scheme of the complex network influencing the uranium (actinide) speciation with focus on plant roots. Reproduced and adapted with permission from Ref. [40]. Copyright (2014) Springer International Publishing is gratefully acknowledged

influence oxidative stress, cell wall and hormone biosynthesis, and signaling pathways (including phosphate signaling). The main actors in iron uptake and signaling (IRT1, FRO2, AHA2, AHA7 and FIT1) were strongly down-regulated. A network calculated using IRT1, FRO2 and FIT1 as bait revealed a set of genes whose expression levels change under uranium stress. The authors present hypotheses to explain how uranium perturbs the iron uptake and signaling response. These results give preliminary insights into the pathways affected by uranium(VI) uptake.

Permission of reproduction by The Royal Society of Chemistry is gratefully acknowledged.

In Fig. 5.2 the suggested uranium interaction with plant cells suggested by Doustaly is shown. The scheme does not include possible uranium redox reactions. However, it can expected that these reactions may play an important role during the uranium uptake by plant cells.

By use of laser-induced spectroscopic methods in cell compartments we could observe absorption and emission spectra which indicate this.

Figure 5.3 shows two photoacoustic spectra 5 and 60 min after addition of 50 μmol/L uranium to the cell growing medium. After filtration of the medium and acidification with H_2SO_4 spectrum of the filtrate was observed. In agreement with the spectroscopic data of pure uranium(IV) and uranium(V) ions we assign the peak at ~645 nm to be a uranium(V) species and the peak at ~662 nm to uranium(IV). This means that a uranium reducing reaction must occur in the medium.

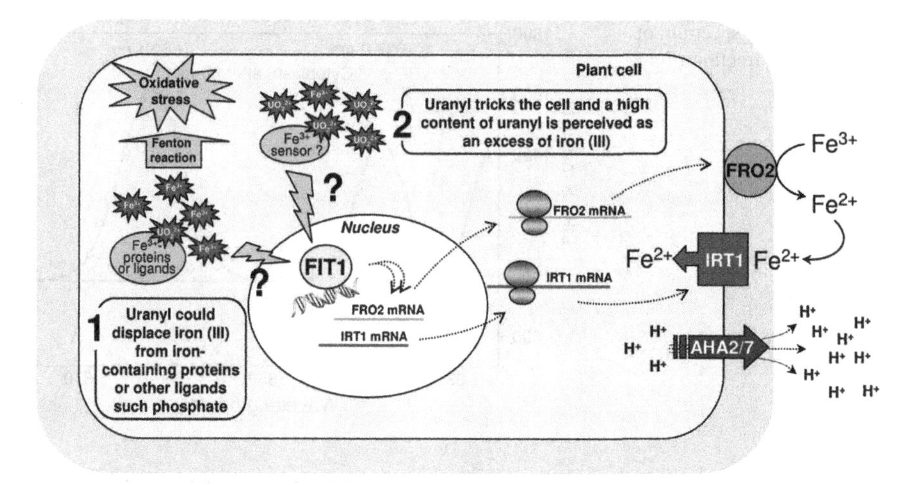

Fig. 5.2 Showing the interaction of uranium with plant cells suggested by Doustaly [42] Permission of reproduction by The Royal Society of Chemistry is gratefully acknowledged.

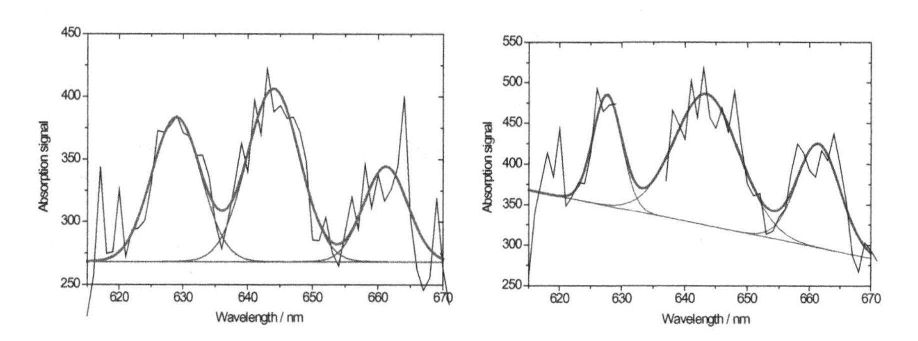

Fig. 5.3 LIPAS spectra of the medium after addition of uranium (5 and 60 min)

The redox potentials of the two reactions are

$$Fe^{3+} + e^- \rightleftarrows Fe^{2+} E_0 = +0.77\,V$$
$$UO_2^{2+} + e^- \rightleftarrows UO_2^{+} E_0 = +0.16\,V \qquad][$$
$$UO_2^{2} + e^- + 4H^+ \rightleftarrows U^{4+} E_0 = +0.27\,V$$

The given values are relevant for acidic solution, which is not used in the experiments.

After formation of uranium(V), a disproportionation reaction usually occurs:

$$2UO_2^{+} + H^+ \rightleftarrows UOOH^+ + UO_2^{2+}$$

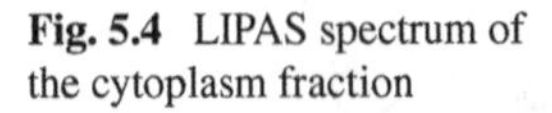

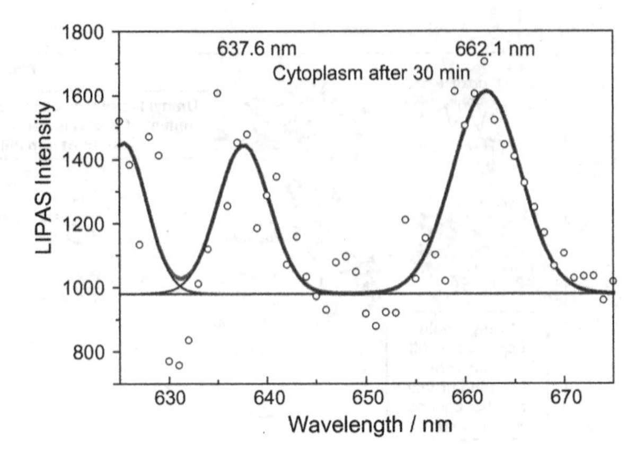

Fig. 5.4 LIPAS spectrum of the cytoplasm fraction

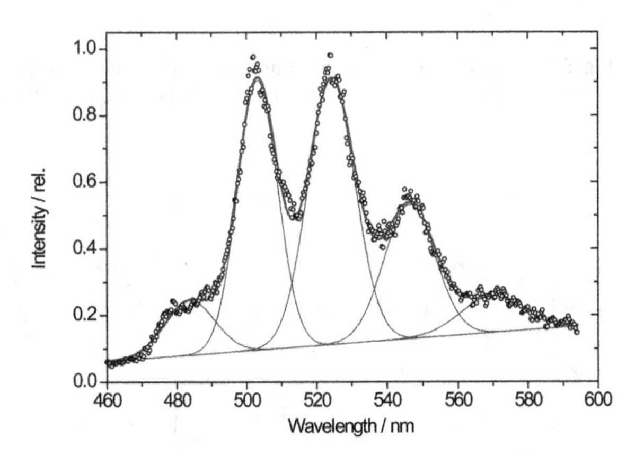

Fig. 5.5 Cryo-luminescence spectrum of the cytoplasm fraction

Therefore, the mechanism reduction of uranium(VI) to uranium(V) at the ferric reduction oxidase FRO2 is most likely. After formation of uranium(V) the iron-regulated transporter FRT1 may than be responsible for the transport of uranium inside the cell. Despite the reducing environment inside cell the disproportionation of uranium(V) inside the cell may continue. Therefore, one should observe uranium in all oxidation states also in cell compartments.

Figure 5.4 shows the absorption spectrum of an acidified cytoplasm fraction. Again we observe the two absorption bands for uranium(V) at 638 nm and for uranium(IV) at 662 nm. By cryo TRLFS measurements in the cytoplasm fraction uranium(VI) could be also detected (Fig. 5.5).

In summary, all three relevant oxidation states of uranium were detected outside and inside the plant cell. This leads to the conclusion that uranium(VI) may be reduced the FRO2, then transported as uranium(V) by IRT1 and disproportionation reactions occur in both compartments. Figure 5.6 gives a simplified model for this, based on the suggestion by Doustaly [42].

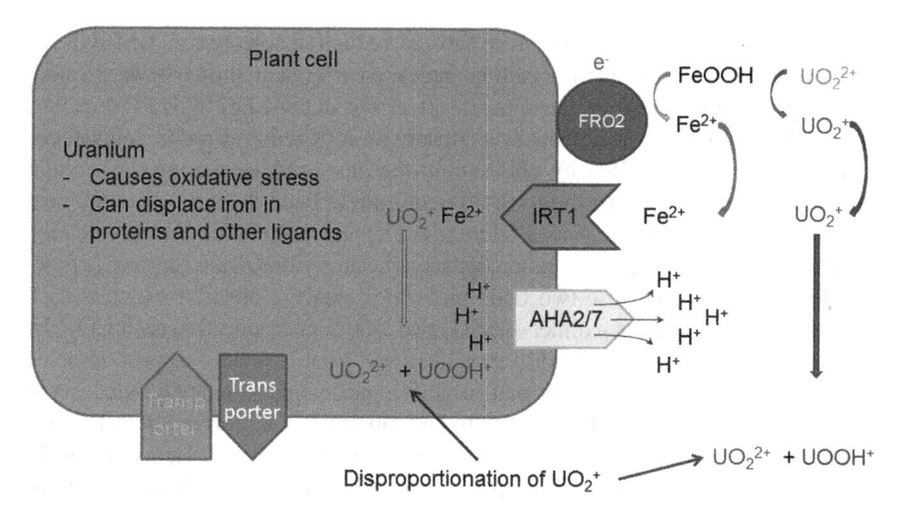

Fig. 5.6 Suggestion for the mechanism of uranium uptake including redox processes. Reference and permission see also Fig. 5.2

In the scheme in Fig. 5.6 also signs for other transporters are included. This may be due to the fact, that it cannot be excluded that uranium may also be transported by other transporters like calcium transporters. Calcium can be easily exchanged by uranium due to nearly same atomic radii.

Hayek and coworkers [43] found that the uranium uptake be *B. juncea* was strongly influenced by calcium in carbonate containing water at circumneutral conditions and at environmental uranium concentrations. This effect was associated with the uranium speciation under these conditions. Especially uranyl- carbonate and the ternary calcium-uranyl-carbonate complex may play an important role. Presence of uranyl-carbonate complexes lead to higher uptake. In contrast to this finding was observed that the uranium uptake in the presence of calcium and carbonate decreases. The formation of the ternary calcium-uranyl-carbonate complex may cause this behavior. In addition, the formation of mineral precipitates on root cell walls was observed. These mineral phases are composed of potassium-uranium-phosphates. The formation of ternary minerals from weathering processes of depleted uranium (DU) has been stated be Schimmack [44]. They found the formation of Sabugalite, an aluminum containing uranium phosphate mineral on sites where DU ammunition has been fired. An explanation where the phosphate ions come from to form this mineral could not be given in this article. As the ammunition corrodes in the nature it seems possible that the phosphate also comes from plant roots.

Uranium as a redox-active heavy metal can cause various redox imbalances in plant cells. Measurements of the cellular glutathione/glutathione disulfide (GSH/GSSG) were performed by Viehweger [34] by HPLC after cellular U contact revealed an interference with this essential redox couple. The GSH content remained unaffected by 10 μM U whereas the GSSG level immediately increased. In contrast, higher U concentrations (50 μM) drastically raised both forms. Using the Nernst equation, it was possible to calculate the half-cell reduction potential of 2GSH/GSSG. In case of lower U contents the cellular redox environment shifted towards more oxidizing conditions whereas the opposite effect was obtained by higher U contents. This indicates that U contact causes a consumption of reduced redox equivalents. Artificial depletion of GSH by chlorodinitrobenzene and measuring the cellular reducing capacity by tetrazolium salt reduction underlined the strong requirement of reduced redox equivalents. An additional element of cellular U detoxification mechanisms is the complex formation between the heavy metal and carboxylic functionalities of GSH. Because two GSH molecules catalyze electron transfers each with one electron forming a dimer (GSSG) two UO_2^{2+} are reduced to each UO_2^{+} by unbound redox sensitive sulfhydryl moieties. UO_2^{+} subsequently disproportionate to UO_2^{2+} and U^{4+}. This explains that in vitro experiments revealed a reduction to uranium(IV) of only around 33% of initial uranium(VI). Cellular uranium(IV) was transiently detected with the highest level after 2 h of U contact. Hence, it can be proposed that these reducing processes are an important element of defense reactions induced by this heavy metal.

Eighteen-day-old *Arabidopsis thaliana* seedlings were exposed to 50 μM uranium. The goal was to investigate the influence of uranium on photosynthetic parameters. A high uranium uptake by roots (50,352 ± 3383 $\mu g\ g^{-1}$ DW at 96 h) was found by Vanhoudt [45]. After uranium contact the plants did not grow anymore. In leaves uranium concentrations have been found to be low (15.0 ± 4.0 $\mu g\ g^{-1}$ DW at 96 h). It could be observed, that the photosynthetic mechanism responded to this uranium stress. The authors concluded from chlorophyll fluorescence measurements and rapid light curves (RLC), that plant leaves start increasing their photosynthetic efficiency and decreasing their non-photochemical quenching under uranium stress. The photosynthetic apparatus shows an optimization at pH 4.5 and seems to be not influenced at pH 7.5 [12].

In an earlier complex study Vanhoudt [46–48] investigated the oxidative stress in *Arabidopsis thaliana.* They treated *Arabidopsis thaliana* with uranium concentrations ranging between 0.1 and 100 μM for 1, 3 and 7 days. Vanhoudt and coworkers studied the reactive oxygen species production and scavenging enzymes at protein and transcriptional level as well as the ascorbate–glutathione cycle under uranium stress. For roots and leaves results were reported. Oxidative stress related responses in the roots were only observed after exposure with 100 μM uranium. Enhancement of lipoxygenase (LOX1) and respiratory burst oxidase homolog (RBOHD) transcript levels could be observed already after the short period of 1 day. In addition, superoxide dismutase (SOD) started during day 1. A correlation between the enhanced SOD-capacity and an enhancement of the expression of iron SOD (FSD1)was found. The catalase (CAT1) transcript levels increase for the detoxification of H_2O_2 simultaneously. In contrast to this, peroxidase capacities were enhanced after 3 days. Additionally, the ascorbate peroxidase capacity and gene expression (APX1) increase. The ascorbate/dehydroascorbate redox couple was shifted towards the oxidized form. This disrupted balance could not be trapped by the glutathione part of the cycle. This cycle was not able to stop the misbalance. Visible responses were found for leaves already after 1 day of exposure. However, uranium concentrations in the leaves were negligible. This suggests a root-to-shoot signaling system. The lipid peroxidation was observed only after exposure with 100 μM uranium. Also the membrane structure and function was affected. Results of lipoxygenase (LOX2) and antioxidative enzyme transcript levels support the transient character of uranium stress responses in leaves. Glutathione concentrations as well as enzyme capacities change with time and concentration, respectively. The ascorbate redox balance in the leaves has been found to be an important modulator of uranium stress. An increase of the total ascorbate concentration and ascorbate/dehydroascorbate redox balance has been found. This depends both on concentration and time.

Hydroxide uranium species were contacted with plant cells (canola). After 24 h contact time the cells were fractionated and the uranium speciation in the fraction was determined by time resolved laser-induced fluorescence spectroscopy at room temperature as well at 150 K. It could be shown that the uranium speciation in the fractions is different to that in the nutrient solution. Also, slightly different spectra were observed at ambient and low temperatures, respectively. As uranium species differ in the luminescence quantum yield the conclusion can be drawn that more than one uranium species may exist in several cell compartments. Comparison of the emission bands with literature data allows assignment of the uranium binding forms. Mainly carboxylic groups are responsible for the uranium binding. However, it could not be excluded that also other groups ($-NH_2^-$; RPO_4^{2-}) may interact with the uranium [49].

Common duckweed (*Lemna minor* L.) is ideally suited to test the impact of metals on freshwater vascular plants. Uranium should cause concentration-dependent oxidative stress and growth retardation on *L. minor*. Horemans [50] used a standardized 7-day growth inhibition test. The uranium impact on *L. minor* growth could be confirmed (EC50 29.5 ± 1.9 μM, EC10 6.5 ± 0.9 μM). The metal-induced oxidative stress response was compared through assessing the activity of different antioxidative enzymes [catalase, glutathione reductase, superoxide dismutase (SOD), ascorbate peroxidase (APOD), guaiacol peroxidase (GPOD) and syringaldizyne peroxidase (SPOD)]. Most of antioxidative enzymes showed significant changes, indicating their importance in counteracting the uranium imposed oxidative burden. Uranium at concentrations below 10 μM increased the level of chlorophyll a and b and carotenoids.

Laurette and coworkers [51] evaluated oilseed rape and sunflower responses to various speciation of uranium. ICP-MS elemental analysis, scanning electron microscopy (SEM) coupled with energy dispersive spectroscopy (EDS), transmission electron microscopy (TEM) and particle-induced X-ray emission spectroscopy (PIXE) were used to study the plant behavior. The response to U follows three schemes. When exposed with the aqua ion UO_2^{2+} the root adsorption and/or accumulation is high, but U transfer to the shoots is limited due to precipitation on cell walls. Treatment with carbonate or citrate reduces U content in roots but a strong increase in shoots was found and uranium was concentrated in leaves. Complex formation with phosphate reduces uranium accumulation in all plant tissues. The uranium was precipitated and adsorbed in clusters on root epidermal cells. Also, the formation of uranium bearing proteins could be confirmed.

Uranium concentrations of 8.65×10^{-3} μg g^{-1} in leaf tissue and 7.95×10^{-3} μg g^{-1} in stem tissue of blueberry plants were reported by Morton [52].

The effect of on uptake and translocation of uranium was followed in *Nicotiana tabacum* L seedlings were used to get information on the influence of pH, phosphates, organic acids (citrate, tartrate, oxalate) and polyamines on uptake and transport of uranium [53]. Leaves and seeds did not show high uranium contents. The uptake and transfer of uranium from the roots to the shoots increases if phosphates are absent. Addition of tartaric acid increases the transport. Higher uptake of uranium occur at pH 3.5 (acetic acid), but transport was lower. Spraying with polyamines (putrescine, cadaverine, spermine and spermidine) influence the uranium uptake in the leaves and the roots negatively.

The uptake of uranium by algae was studied by Fortin [54]. They reported a intracellular absorption. After already 30 min the intracellular absorption has the same concentration dependence as adsorption in the extracellular sphere. The uptake was influenced by citrate and EDTA. Both anions lead to a decrease of the uptake. Also, phosphate ions should influence the uptake, however, the published diagram for addition of phosphate leads to the conclusion, that phosphate had only small or no influence. The intracellular uranium concentration is strongly correlated to the free uranyl ion in solution. Free uranyl ions should dominate the mechanism of the cellular uptake. This suggestion coincides with the results made by Doustaly [42] and also the results from laser-induced spectroscopic studies, as shown above.

For green algae *Chlorella vulgaris* has been studied uranium biosorption and possible intracellular accumulation of this metal has been studied [55]. Release of intracellular compounds like organic acids during the first 96 h of contact with uranium has been investigated. Identification of the metal speciation were tried by TRLFS, extended X-ray absorption fine structure (EXAFS) spectroscopy as well as attenuated total reflection Fourier transform infrared (ATR-FTIR) spectroscopy. The applied experimental conditions strongly influence the coordination of uranium. As important result the authors concluded, that the coordination via carboxylic groups is the preferred binding form of the metal. However, it can be also stated, that fractionation of cellular compartments prior analysis using the above mentioned methods is needed.

Uranyl nitrate (10 μM) was added to suspension cultures of canola (*Brassica napus*). After 24 h the cells were disrupted and then fractionated. The applied protocol was published by Larsson and co-workers [56]. The uranium speciation was tried to be determined by TRLFS in the several cell fractions (cytoplasm fraction, plasma membrane fraction and cell nuclei fraction). TRLFS spectra were measured with two setups, first at room temperature and shortly after this by a cryogenic setup at 150 K. It has been stated already that uranium shows a hot emission band. This makes the comparison of spectra recorded under cryogenic conditions with those, which were recorded at room temperature somewhat tricky. This is due to the fact that from the hot band as well as from lower emission level transitions to the vibronic levels in the ground state occur. Due to the fact that the transitions of both emitting level overlap, a clear deconvolution of the spectra often is not possible. With decreasing temperature the transitions from the hot band disappear. This leads in some cases to

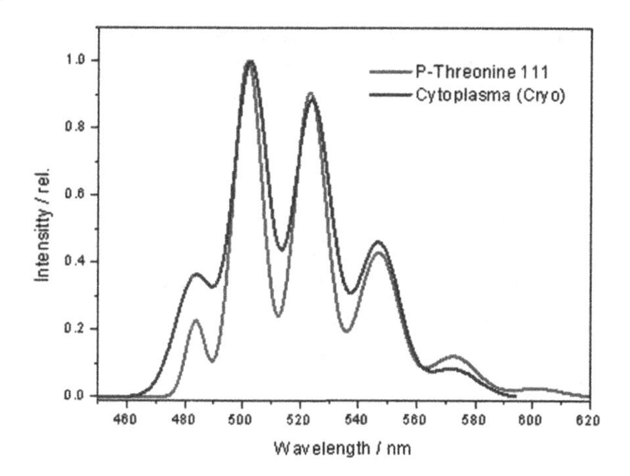

Fig. 5.7 Luminescence spectrum of uranium(VI) in the cytoplasm fraction at 150 K and the uranium-phospho-threonine complex. Reproduced with permission from Ref. [40]. Copyright (2014) Springer International Publishing is gratefully acknowledged

a small but remarkable shift of the band maxima from the lower emission level at cryogenic temperatures.

The luminescence spectrum at 150 K of the cytoplasm fraction in comparison the uranium phospho-threonine complex is shown in Fig. 5.7. The assignment of the measured spectra by comparison of these spectra with databases is mostly a complicated process. In the case of cytoplasm, the spectrum is comparable to spectra, which were obtained for uranium(VI)—ortho-Phospho-L-Threonine [57].

However, for the other fractions the assignment is mostly more complicated. Besides the agreement of the localization maxima of the emission bands other factors derived from the spectra as the intensity distribution between the several transitions and the full width at half maximum of an emission are important data. Last not least also the luminescence lifetime should to be taken into consideration. However, the luminescence lifetime may be influenced by several additional factors like composition of the measured sample (solution, measurement, temperature and others). Especially the temperature of the sample during the measurement influences the luminescence lifetime strongly due to changing the dynamic quench processes, which play an important role in uranium luminescence spectroscopy.

Table 5.1 summarizes the data from the luminescence measurements. For comparison the data for uranium and several uranium complexes are added.

The spectra recorded at room temperature and under cryogenic conditions are shown in Fig. 10. Clearly the difference in the spectra from the membrane and the residual solid fraction can be derived. However, also slight differences between room temperature and cryogenic measurement become visible. In addition from the noise in the measurement of the membrane fraction under cryogenic conditions it can be concluded, that freezing of the sample leads not in any case to better spectra. Despite the discussed decrease of the hot band and the decrease of dynamic quench processes due to the freezing, another luminescence property has to be taken into account. Different species of the uranyl ion show different luminescence quantum

Table 5.1 Emission maxima obtained for cell compartments and selected uranium compounds (RT—room temperature; Cr—Cryogenic measurement)

	Emission wavelength (nm)						References
Cell compartments							
Cytoplasma (RT)	481.2	498.7	520.2	543.8	568.4		
Cytoplasma (Cr)	483.7	503.2	524.3	546.3	569.7		
Residual solids (RT)	481.6	498.1	518.8	541.0	564.1		
Residual solids (Cr)	482.1	503.4	523.7	545.2	571.3		
Membrane (RT)	483.4	499.0	519.3	539.7	557.1		
Membrane (Cr)	483.1	504.1	522.4	542.1	562.7		
Reference compounds							
UO_2^{2+}	472.0	488.7	509.7	533.5	559.4	585.4	
$(UO_2)_3(OH)_5^+$	484.0	497.0	513.5	534.7	556.6	578.8	[58]
$UO_2(CO_3)_3^{4-}$	480.5	498.6	519.3	541.6	567.3		
$Ca_2UO_2(CO_3)_3$	484.6	504.1	525.4	549.1	573.4		[59]
ATP	480.9	495.3	516.6	540.2	564.6	593.8	[60]
O-Phospho-L-Threonine 111	483.7	501.8	523.4	546.9	572.7	601.6	[57]
O-Phospho-L-Threonine 112	479.2	495.6	517.3	541.0	565.7	594.5	[57]
O-Phospho-L-Threonine 113	477.8	493.9	515.1	539.5	563.9	593.3	[57]
Phospholipids	480.7	497.6	519.4	543.0	568.8	598.3	[61]
Glycine	478.7	495.3	516.6	540.6	565.0	594.4	[62]
Malonate	477.1	493.6	514.6	537.7	563.5	590.1	[63]
Citric acid 11	474.4	491.2	513.4	537.3	561.5		[64]
Citric acid 12	483.7	502.7	524.5	548.1	573.9		[64]
Fructose-6-phosphate	482.0	496.0	516.5	540.1	565.1	593.8	[65]

yields. This may also lead to different shapes of the spectra as by freezing other species may become more dominant in the spectrum.

For the spectra shown in Fig. 5.8 detailed data are listed in Table 5.2.

Nevertheless, this comparison gives only a first hint to the uranium speciation in the cytoplasm fraction of the used plant cells. Other species cannot be excluded. For this more detailed information will be necessary. By comparing the data in Table 5.2, a difference in emission maxima up to 1.4 nm has to be taken into account. The FWHM shows also does show also differences up to 8.9 nm. This is also the case for the area fit. However, for the three main emission bands the agreement is relatively good. Another problem of the measurement of fractionated cells may occur during the disruption of tissues and cells. It cannot be excluded that during this process the metal speciation is changed, if other ligands are present which have higher complex formation constants.

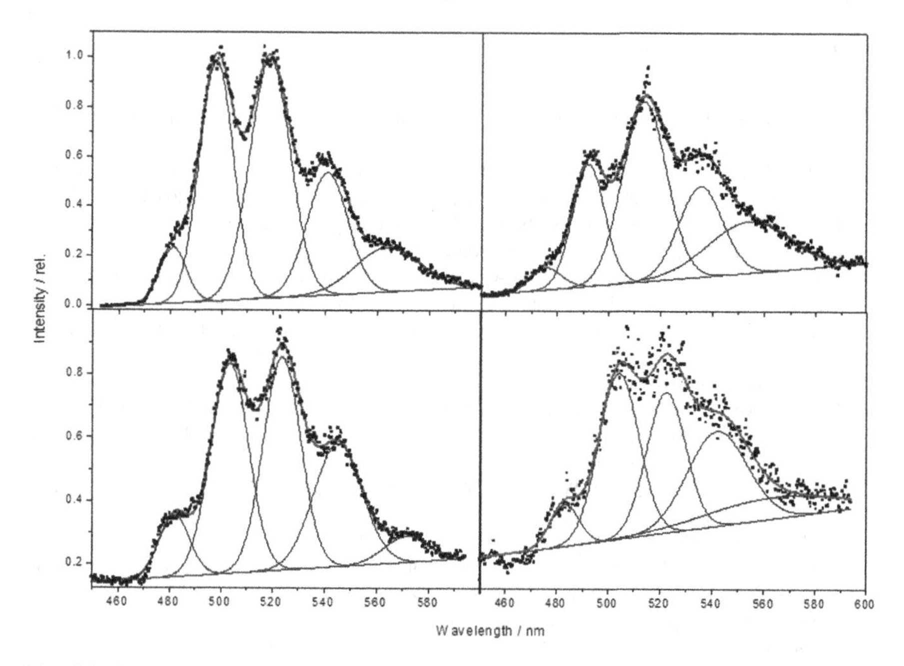

Fig. 5.8 Luminescence spectra for cell compartments (left—residual solid fraction, right—membrane fraction, top—room temperature measurement, bottom—cryogenic measurement)

Table 5.2 Spectral data of uranium(VI) species in the cytoplasm fraction and in uranium(VI)-ortho-phospho-L-threonine-complex

Cytoplasm fraction (Cryogenic data)			Uranium(VI)-ortho-phospho-L-Threonine		
Emission maximum (nm)	Full width at half maximum (Nm)	Area (%)	Emission maximum nm	Full width at half maximum (nm)	Area (%)
483.7	16.8	7.8	483.7	7.9	5.0
503.2	13.6	30.9	501.8	12.0	33.8
524.3	15.8	35.5	523.4	13.1	33.4
546.3	17.0	19.3	546.9	16.3	19.7
569.7	20.8	6.5	572.7	19.0	6.5

References

1. Ernst WHO (1996) Biovailability of heavy metals and decontamination of soils by plants. Apl Geochem 11:163–167
2. Ehlken S, Kirchner G (2002) Environmental processes affecting plant root uptake of radioactive trace elements and variability of transfer factor data: a review. J Environ Radioactiv 58:97–112
3. Ribera D, Labrot F, Tisnerat G, Narbonne JF (1996) Uranium in the environment: occurrence, transfer, and biological effects. Rev Environ Contam T 146:53–89
4. Perez-Sanchez D, Thorne MC (2014) An investigation into the upward transport or uranium-series radionuclides in soils and uptake by plants. J Radiol Prot 34:545–573
5. Rufyikiri G, Thiry Y, Declerck S (2003) Contribution of hyphae and roots to uranium uptake and translocation by arbuscular mycorrhizal carrot roots under root-organ culture conditions. New Phytol 158:391–399
6. Vandenhove H, Van Hees M, Wouters K, Wannijn J (2007) Can we predict uranium bioavailability based on soil parameters? Part 1: effect of soil parameters on soil solution uranium concentration. Environ Pollut 145:587–595
7. Vandenhove H, Van Hees M, Wouters K, Wannijn J, Wang L (2007) Can we predict uranium bioavailability based on soil parameters? Part 2: Soil solution uranium concentration is not a good bioavailability index. Environ Pollut 145:577–586
8. Caldwell EF, Duff MC, Ferguson CE, Coughlin DP, Hicks RA, Dixon E (2012) Bio-monitoring for uranium using stream-side terrestrial plants and macrophytes. J Environ Monit 14:968–976
9. Viehweger K, Geipel G (2010) Uranium accumulation and tolerance in *Arabidopsis halleri* under native versus hydroponic conditions. Environ Exp Bot 69:39–46
10. Aranjuelo I, Doustaly F, Cela J, Porcel R, Mueller M, Aroca R, Munne-Bosch S, Bourguignon J (2014) Glutathione and transpiration as key factors conditioning oxidative stress in *Arabidopsis thaliana* exposed to uranium. Planta 239:817–830
11. Haas JR, Bailay EH, Purvis OW (1998) Bioaccumulation of metals b lichens: uptake of aqueous uranium by *Peltigera membranacea* as a function of time and pH. Am Mineral 83:1494–1502
12. Markich SJ (2002) Uranium speciation and bioavailability in aquatic systems: an overview. TheScientificWorldJ 2:707–729
13. Sheppard SC, Evenden WG (1988) Critical compilation and review of plat/soil concentration ratios for uranium, thorium and lead. J Environ Radioactiv 8:255–285
14. Kaur A, Singh S, Virk HS (1988) A study of uranium uptake in plants. Nucl Tracks Rad Meas 15:795–798
15. Singh KP (1997) Uranium uptake by plants. Curr Sci India 73:532–535
16. Frindik O (1986) Uranium contents in soils, plants and foods. Landwirt Forsch 39:75–86
17. Nalezinski S, Lux D (1998) in BfS annual report 1997, 37–42
18. Ibrahim SA, Whicker FW (1992) Comparative plant uptake and environmental behavior of U-series radionuclides at a uranium mine-mill. J Radioanal Nucl Ch Ar 156:253–267
19. Mitchell N, Perez-Sanchez D, Thorne MC (2013) A review of the behaviour of U-238 series radionuclides in soils and plants. J Radiol Prot 33:R17–R48
20. Yaprak G, Cam NF, Yener G (1998) Determination of uranium in plants frolic high background area by instrumental neutron activation analysis. J Radioanal Nucl Ch Ar 238:167–173
21. Mortvedt JJ (1994) Plant and soil relationships of uranium and thorium decay series radionuclides—a review. J Environ Qual 23:643–648
22. Simon LS, Ibrahim SA (1990) Biological uptake of radium by terrestrial plants. In: The environmental behavior of Radium. Technical reports series 310, Int. Atomic Energy Agency, Vienna, Austria
23. Baker AJM (1981) Accumulators and excluders—strategies in the response of plants to heavy metals. J Plant Nutr 3:643–654
24. Huang JWW, Blaylock MJ, Kapulnik Y et al (1998) Phytoremediation of uranium contaminated soils: role of organic acids in triggering uranium hyperaccumulation in plants. Env Sci Technol 32:2004–2008

25. Joshi-Tope G, Francis AJ (1995) Mechanisms of biodegradation of metal-citrate complexes by *Pseudomonas fluorescens*. J Bacteriol 177:1989–1993
26. Francis AJ, Dodge CJ, Gillow JB (1992) Biodegradation of metal citrate complexes and implications for toxic-metal mobility. Nature 356:140–142
27. Ebbs SD, Brady DJ, Kochian LV (1998) Role of uranium speciation in the uptake and translocation of uranium by plants. J Exp Botany 49:1183–1190
28. Guenther A, Bernhard G, Geipel G et al (2003) Uranium speciation in plants. Radiochim Acta 91:319–328
29. Laurette J, Larue C, Mariet C, Brisset F, Khodja H, Bourguignon J, Carriere M (2012) Influence of uranium speciation on its accumulation and translocation in three plant species: oilseed rape, sunflower and wheat. Environ Exp Bot 77:96–107
30. Rossberg A, Barkleit A, Tsushima S, Scheinost AC, Kaden P, Stumpf T (2017) A multi-method approach for the investigation of complex actinide systems: uranium(VI) interactions with DNA and sugar phosphates. Migration Conference, Barcelona, Spain
31. Cvetkovic A, Menon AL, Thorgersen MP et al (2010) Microbial metalloproteomes are largely uncharacterized. Nature 466:779–782
32. Maret W (2010) Metalloproteomics, metalloproteomes, and the annotation of metalloproteins. Metallomics 2:117–125
33. Yannone SM, Hartung S, Menon AL et al (2012) Metals in biology: defining metalloproteomes. Curr Opin Biotechnol 23:89–95
34. Viehweger K, Geipel G, Bernhard G (2011) Impact of uranium (U) on the cellular glutathione pool and resultant consequences for the redox status of U. Biometals 24:1197–1204
35. Frost L, Geipel G, Viehweger K et al (2011) Interaction of uranium(VI) towards glutathione—an example to study different functional groups in one molecule. Proc Radiochem A Suppl Radiochim Acta 1:357–362
36. Nazir M, Naqvi II (2010) Synthesis and characterization of uranium(IV) complexes with various amino acids. J Saudi Chem Soc 14:101–104
37. Choppin GR (2007) Actinide speciation in the environment. J Radioanal Nucl Ch Ar 273:695–703
38. Kramer U, Talke IN, Hanikenne M (2007) Transition metal transport. FEBS Lett 581:2263–2272
39. Kobayashi T, Nishizawa NK (2012) Iron uptake, translocation, and regulation in higher plants Ann Rev. Plant Biol 63:131–152
40. Geipel G, Viehweger K (2014) Speciation of actinides after plant uptake. In: Gupta DK, Walther C (eds) Radionuclide contamination and remediation through plants. Springer International Publishing, 197–213
41. Viehweger K (2014) How plants cope with heavy metals. Bot Stud 55:35. http://www.as-botanicalstudies.com/content/55/1/35
42. Doustaly F, Combes F, Fievet JB, Berthet S, Hugouvieux V, Bastien O, Aranjuelo I, Leonhardt N, Rivasseau C, Carriere M, Vavasseur A, Renou JP, Vandenbrouck Y, Bourguignon J (2014) Uranium perturbs signaling and iron uptake response in *Arabidopsis thaliana* roots. Metallomics 6:809–821
43. El Hayek E, Torres C, Rodriguez-Freire L, Blake JM, De Vore CL, Brearley AJ, Spilde MN, St. Cabaniss, Ali A-MS, Cerrato JM (2018) Effect of calcium on the bioavailability of dissolved Uranium(VI) in plant roots under circumneutral pH. Environ Sci Technol 52:13089–13098
44. Schimmack W, Gerstmann U, Schultz W, Geipel G (2007) Long-term corrosion and leaching of depleted uranium (DU) in soil. Radiat Environ Biophys 46:221–227
45. Vanhoudt N, Horemans N, Biermans G, Saenen E, Wannijn J, Nauts R, Van Hees M, Vandenhove H (2014) Uranium affects photosynthetic parameters in *Arabidopsis thaliana*. Environ Exp Bot 97:22–29
46. Saenen E, Horemans N, Vanhoudt N, Varidenhove H, Biermans G, Van Hees M, Wannijn J, Vangronsveld J, Cuypers A (2014) The pH strongly influences the uranium-induced effects on the photosynthetic apparatus of *Arabidopsis thaliana* plants. Plant Physiol Bioch 82:254–262

47. Vanhoudt N, Vandenhove H, Horemans N, Remans T, Opdenakker K, Smeets K, Bello DM, Wannijn J, Van Hees M, Vangronsveld J, Cuypers A (2011) Unraveling uranium induced oxidative stress related responses in *Arabidopsis thaliana* seedlings. Part I: responses in the roots J Environ Radioactiv 102:630–637
48. Vanhoudt N, Cuypers A, Horemans N, Remans T, Opdenakker K, Smeets K, Bello DM, Havaux M, Wannijn J, Van Hees M, Vangronsveld J, Vandenhove H (2011) Unraveling uranium induced oxidative stress related responses in *Arabidopsis thaliana* seedlings. Part II: responses in the leaves and general conclusions J Environ Radioactiv 102:638–645
49. Geipel G, Viehweger K (2015) Speciation of uranium in compartments of living cells. Biometals 28:529–539
50. Horemans N, Van Hees M, Van Hoeck A, Saenen E, De Meutter T, Nauts R, Blust R, Vandenhove H (2015) Uranium and cadmium provoke different oxidative stress responses in *Lemna minor* L. Plant Biol 17:91–100
51. Laurette J, Larue C, Llorens I, Jaillard D, Jouneau PH, Bourguignon J, Carriere M (2012) Speciation of uranium in plants upon root accumulation and root-to-shoot translocation: a XAS and TEM study. Environ Exp Bot 77:87–95
52. Morton LS, Evans CV, Estes GO (2002) Natural uranium and thorium distributions in podzolized soils and native blueberry. J Environ Qual 31:155–162
53. Soudek P, Petrova S, Buzek M, Lhotsky O, Vanek T (2014) Uranium uptake in *Nicotiana* sp under hydroponic conditions. J Geochem Explor 142:130–137
54. Fortin C, Dutel L, Garnier-Laplace J (2004) Uranium complexation and uptake by a green algae in relation to chemical speciation: the importance of the free uranyl ion. Env Technol Chem 23:974–981
55. Vogel M, Guenther A, Rossberg A et al (2010) Biosorption of U(VI) by the green algae *Chlorella vulgaris* in dependence of pH value and cell activity. Sci Total Env 409:384–395
56. Larsson C, Sommarin M, Widell S (1994) Isolation of highly purified plant plasma-membranes and separation of inside-out and right-side-out vesicles. Method Enzymol 228:451–469
57. Guenther A, Geipel G, Bernhard G (2006) Complex formation of U(VI) with the amino acid L-threonine and the corresponding phosphate ester O-phospho-L-threonine. Radiochim Acta 94:845–851
58. Sachs S, Brendler V, Geipel G (2006) Uranium(VI) complexation by humic acid under neutral pH conditions studied by laser-induced fluorescence spectroscopy. Radiochim Acta 95:103–110
59. Bernhard G, Geipel G, Reich T, Brendler V, Amayri S, Nitsche H (2001) Uranyl(VI) carbonate complex formation: validation of the $Ca_2UO_2(CO_3)_3$(aq.) species. Radiochim Acta 89:511–518
60. Geipel G, Bernhard G, Brendler V, Reich T (2000) 5th International Conference on Nuclear and Radiochemistry. Extended Abstracts, vol 2, Pontresina, Switzerland, 473–476
61. Koban A, Bernhard G (2007) Uranium(VI) complexes with phospholipid model compounds—a laser spectroscopic study. J Inorg Biochem 101:750–757
62. Günther A, Geipel G, Bernhard G (2007) Complex formation of uranium(VI) with the amino acids L-glycine and L-cysteine: a fluorescence emission and UV-Vis absorption study. Polyhedron 26:59–65
63. Brachmann A, Geipel G, Bernhard G, Nitsche H (2002) Study of uranyl(VI) malonate complexation by time resolved laser-induced fluorescence spectroscopy (TRLFS). Radiochim Acta 90:147–149
64. Guenther A, Steudtner R, Schmeide K et al (2011) Luminescence properties of uranium(VI) citrate and uranium(VI) oxalate species and their application in the determination of complex formation constants. Radiochim Acta 99:535–541
65. Koban A, Geipel G, Roßberg A, Bernhard G (2004) Uranium(VI) complexes with sugar phosphates in aqueous solution. Radiochim Acta 92:903–908

Chapter 6
Outlook

This overview can be only a first sight to the processes and mechanism of uptake and metabolism of uranium as an element in the actinide series by plants. This may be important as plants are part of the food chain. By knowing the uptake processes it may be probably possible to reduce in future the uranium intake of humans.

One possibility to get effort on this way may be the study of interactions of actinides and in special case of uranium with pure ligands with importance to plants species. An example how multiple experimental methods can be used to determine such interactions between uranium and pure components from biosystems has been shown by Kretschmar et al. [1]. The authors used NMR spectroscopy, cryogenic time-resolved luminescence spectroscopy and FT-IR-spectroscopy in combination with modern mathematic methods, like factor analysis to determine the interaction of uranium(VI) with the bioligand. Due to the NMR study D_2O was used as solvent. It could be shown that a binary complex was formed. The stability constant at infinite dilution was determined to be $\log K^K = 5.24 \pm 0.08$ for the reaction

$$UO_2^{2+} + H_3GSSG^- \rightleftarrows UO_2(H_3GGSG)^+$$

The study showed that in whole range of pD (2–8) a precipitate was formed. The solubility was least for $4 < pD < 6.5$. A speciation diagram shows that in the pH region between pH 3 and 4 the amount of complex formation reaches nearly 40%. This may be one possibility to achieve more detailed information about interaction of uranium with component parts of living cells.

A way to study the interaction of uranium with cells in suspension culture has been examined by Rajabi [2]. A multi-analytical approach was applied to study these interactions. Biochemical analysis, spectroscopic tools and thermodynamic modelling were combined. The study was focused on effects of uranium on cell viability and the homeostasis of essential nutrient constituents. The speciation of uranium was found to play an important role under Fe-sufficient and deficient conditions and connected to this in uranium toxicity. The uranium toxicity threshold is affected strongly on the physiological characteristics of the plant species and their

G. Geipel, *Uranium and Plant Metabolism*, SpringerBriefs in Biometals,
https://doi.org/10.1007/978-3-030-80815-0_6

growth conditions. Early presence of uranium in the nutrient solution indicate the initiation of a cellular defence mechanism. It was concluded that Ca channels are involved in this mechanism. Uranium and iron show synergistic toxic effects and this indicates that interactions among different metal ions influence their toxicity.

Nevertheless, in the near future also uptake processes of higher actinides may be of interest. Here especially the element neptunium, plutonium and americium can become important. This is due to the nuclear accidents in the past (Chernobyl and Fukushima) as well as discussions about dismantling nuclear power stations and the nuclear waste storage in deep underground. However, these neptunium and plutonium do not show luminescence properties and therefore the observation of these elements in plant cells may be much more difficult. One way to study higher actinides may consist in the investigation of curium or, if radioactive elements are not useable, lanthanides.

In a study, plant cells of *Brassica napus* were treated with uranium and europium [3]. Both metal ions were mainly enriched in the cell wall fraction. The authors conclude that this behaviour is caused by effective mechanism of protection of the living cell. However, under extreme metal stress both metals can also be detected in the cytosol, indicating uptake of the metal species. In addition to this an influence on the Ca(II) homeostasis was observed. Analysis of the obtained luminescence spectra confirmed that the plant cells provided multiple binding environments for both metal ions. For europium two different inner-sphere species were detected. In case of uranium inorganic and organic phosphate groups dominate the binding forms.

Europium and curium are toxic metal ions and were used as luminescence probes for plant cell interactions by Moll [4]. The plant cell response of *Brassica napus* to treatment with europium(III) and curium(III) was studies by luminescence measurements. The experiments were carried out over a growth period of seven days. In both series a bio-association of the toxic metal was observed. The plant cells contained 0.58 μmol Eu/$g_{\text{fresh cells}}$ and 1.82 μmol Cm/$g_{\text{fresh cells}}$ after treatment with 30 μM Eu(III)and 0.68 μM Cm(III), respectively. In case of curium a biosorption process was identified after 5 h of contact with the metal ion. 73 4% of the added curium were found to be associated to the plant cells. UV excitation and site sensitive excitation lead to the conclusion that the Eu(III) coordinates in carboxyl and organic phosphate groups, whereas in case of curium from spectroscopic data were concluded that protein-based carboxyl group are involved. Contribution of phosphate groups may also possible.

Using luminescent metal ions as probes for the study of the interaction with plant cells provide wide possibilities to obtain insights in phenomena of uptake, storage and transfer of these metals. Important results may be expected in future also towards toxicity of metal ions.

References

1. Kretschmar J, Strobel A, Haubitz T et al (2020) Uranium(VI) complexes of glutathione disulfide forming in aqueous solution. Inorg Chem 59:4244–4254
2. Rajabi F, Jessat J, Garimella JN et al (2021) Uranium(VI) toxicity in *tobacco BY-2* cell suspension culture - a physiological study. Ecotox Env Saf 211:111883. https://doi.org/10.1016/j.ecoenv.2020.111883
3. Moll H, Sachs S, Geipel G (2020) Plant cell (*Brassica napus*) response to europium(III) and uranium(VI) exposure. Environ Sci Pollut Res 25:32048–32061. https://doi.org/10.1007/s11356-020-09525-2
4. Moll H, Schmidt M, Sachs S (2021) Curium(III) and europium(III) as luminescence probes for plant cell (*Brassica napus*) interactions with potentially toxic metals. J Haz Mat. 412:125251, https://doi.org/10.1016/j.jhazmat.2021.125251

Printed in the United States
by Baker & Taylor Publisher Services